The Tao Of Martial Business

2 = 250K

By Mike & Andrea Patterson

ISBN-10: 098585572X
ISBN-13: 978-0-9858557-2-7

The Tao of Martial Business - If you can't afford it, you REALLY need it!
Making a Fortune in the "Adults" Martial Arts Marketplace

Would you and your partner (senior student, or hired employee) like to generate two hundred and fifty thousand dollars net profit per year as professional martial arts instructors?

Are you a part-time instructor who also has a full-time job just to make ends meet?

You can realize the dream of being a master instructor and live the life style of afforded by a six figure annual income. It *IS* possible to achieve this goal without becoming a franchise of some martial arts "fast food chain" structure. You do not have to pay costly annual franchise fees. You do not have to fly the flags of other organizations just to learn their secrets, which may or may not be applicable to your unique kung fu style. You *CAN* promote your own kung fu brand, *KEEP* the pride and integrity of your arts and make your family martial arts business a viable source of income! You can learn to be *PROFITABLE* in this highly lucrative business!

Forward

When I was young, I had an aversion to money. I had been raised to believe in the perceived truth of so called "principles" like "money is the root of all evil" which only much later did I discover was an innocent, yet in my case damning, misquote and should actually have been taught to me as "the LOVE of money is the root of all evil". That three word addition makes a big difference in interpretation and the subsequent adoption of an individual value associated with that statement.

Because such thoughts were placed into a very young awareness and never fully explained or even correctly transmitted, my mind began to skew itself toward a direction of financial self sabotage whenever I began to be even a little successful.

In addition, through all prevalent media in those days, Kung Fu masters were always portrayed as austere and aesthetic in their pursuits. They were often painted as financially destitute. To many, these so called "values" are still representative of what is thought to be a "real" master teacher even today. And it really could not be further from the truth.

This is a mindset that, if left unchecked, will prevent success in the modern world. It is a fictitious and romanticized notion that has been constructed for the "movie sifu" because such an image is far easier to sell in the movie medium than is the truth. And the truth is so much less appealing as far as Hollywood is concerned. The real truth being that skilled professionals, no matter from what area of expertise, have always been fairly compensated for the skills that they possess. And so they should be.

In elder times, if a Kung Fu master truly had skill, there were several options available in terms of pursuing a viable financial avenue for economic security. Gainful employment as a compound teacher to a wealthy family, government muscle, personal bodyguard or caravan guard to name a few. Some of these avenues still exist today. But the short of the story is that these elder Sifu were not financially destitute by any means. They were working professionals and made a good living, just as you should consider attaining for yourself in today's venues.

You have worked long and hard to achieve the skills you possess and can now consequently offer to your local marketplace. The time you have spent in acquiring your knowledge base is equivalent in time spent to acquire any college degree and then some.

An educated professional charges a premium for the knowledge acquired and skills to implement that same knowledge in their respective marketplaces. You are no different and should understand, and take to heart, that you are well worth what you charge. Besides which, if you do not charge a fair premium for your services, several things could begin to evolve that would be detrimental to your students and their well being.

Charging a fair premium means being able to pay the rent so that your students have a place to train. It means being able to heat and air condition the facility so that those same students may have a comfortable environment in which to train. It means being able to purchase, upgrade and keep in good working condition all the specialized equipment students would like to have in their school. It means being able to provide a first grade experience for them while they engage in their studies. And it means that their Sifu will be less stressed about earning a living so that he/she can devote their entire minds to giving the best instruction and martial education they are capable of providing.

Besides which, if you fail in your teaching endeavor who will you then be able to serve? You must provide an equitable living for yourself before you have the ability to assist others in their needs and endeavors.

Statement of Purpose

We have written this manual for you, the aspiring professional martial arts teacher. The person who wishes to turn passion into profit. To make your avocation your vocation. To give you the freedom to do what you love and make a good living while doing exactly that.

Drawn from our experiences as lifetime professional martial arts teachers, we have compiled a "no B.S." straight talk manual to assist you in realizing your dreams. This manual will allow you to be an adult, teaching adults. It will allow you to take your passion and direct it in such a way that you will help thousands with their own dreams, goals and aspirations over the next several decades as a professional martial arts teacher.

This manual will help you to help yourself, your community and your pupils to become a more productive and disciplined group of people, one student by cared for student, at a time.

Always remember: It is not about you. It is about the students. We are there to serve them.

A bit about your authors...

MASTER MIKE PATTERSON:

44 years of Martial Arts experience (at time of this writing)
Founder - Innovative Program R.S.P.C.T. (realistically structured progressive combat training)
Highly Successful Owner/Operator - Adult's Only Chinese Martial Arts Schools
West Coast Regional Director - United States Chinese KuoShu Federation
Special advisor - United States Chinese KuoShu Federation
Men's Coach - World Team U.S.A.
Coach Of The Year - Inside Kung Fu Magazine
Consultant for military, law enforcement and professional athletics programs
Publisher - International Martial Arts Magazine
Producer - full line of Internationally recognized Instructional Videos
Founder/Author of the high profile website hsing-i.com

MASTER ANDREA CHENG:

29 years of Martial Arts experience (at time of this writing)
Million Dollar Manager - Fortune 500 Company
Certified Training Manager - Fortune 500 Company
Top 2% Performer (national) - Fortune 500 Company
Fiscal Performance Troubleshooter - Fortune 500 Company
Internationally Certified Kuoshu Referee
Internationally Certified Internal Martial Arts Judge
Overall Team Leader - World Team U.S.A.
Consultant - personal training programs for business professionals
Chief Journalist - for International Martial Arts Magazine
Consultant - International Martial Arts Magazine

***** Note: Specific case results may vary. The authors are not responsible for any individual's or group's inability to utilize effectively the strategies presented in this book. Results may vary according to specific markets, locales, individual perception and other mitigating factors which are beyond our control.**

***** Note: Specific case results may vary. The authors are not responsible for any individual's or group's inability to utilize effectively the strategies presented in this book. Results may vary according to specific markets, locales, individual perception and other mitigating factors which are beyond our control.**

Scouting a Location

This is probably one of the more involved, yet underappreciated aspects of business. We all know the saying; "location, location, location". But do we really comprehend what exactly that means? Read on to make sure you understand this aspect.

What is your Market?

Before you do anything at all in terms of scouting a potential location you need to decide what your market is going to be. I know, I know.. All those who study martial arts right? Wrong. You cannot capture ALL people. You must decide what kind of people YOU WANT in your programs and go after *that* market share specifically.

For example: Do you really want Joe Gangbanger and his cronies in your programs? If the answer is no (which is highly recommended) then you must do three things. One, create programs that will not likely appeal to that sort of an individual. Two, price your programs beyond his reach. And three, *locate* your school in an area that is not predominantly habited by such individuals.

If, on the other hand, you DO want this sort of person in your programs, then price them accordingly and create hard-core, blood and guts programs that will appeal to his inner beast. Locate your school in the middle of his territory. Your school will be filled, you will draw the focus of all law enforcement professionals (not in a good way) and you will make no money. But hey, you're running an ego driven, hardcore program and you're happy right?

So you see, the type of market you are trying to procure may dictate location options at your disposal. If you wish to garner the professional individual, those people will not want to park their BMWs, Mercedes, etc. in a seedy area of town. These perspectives MUST be factored into your location search process.

Do your Demographics Due Diligence.

Statistically, only 2% of the populace is actually interested in studying any martial art. You need to determine if that 2% is available in the area of your interest before you pour your money and time into the effort of finding a location.

Go to your local City Chamber of Commerce and pull the demographics for the areas of your city in which you are interested in establishing yourself. Find out the key statistics of the area.

Population Density - A greater density is better than lesser density as most people are only willing to drive a few miles for any type of service they seek, including martial arts.

Types Of Households - In example, Single family vs. Condominiums, as single established family households will tend to have more disposable income.

Household Income - The bigger the average number the better, as once again these households can more likely afford your services.

Average age - This is your specific market target and you must decide ahead of time who you want to teach mostly in terms of age group.

Education Level - Again, who do you want to teach?

Crime Rate - High crime generally means less students. Unlike in the movies, good people in high crime areas do not enroll in your school to combat crime. Good people MOVE.

Performing your demographic research will pay big dividends in helping you select a quality location with sustainable long term growth. You don't want to put in all the energy required for starting, and then beginning to grow your school, only to discover that you need to change your location a year from now.

Disposable income is key.

There is generally more disposable income in single family households in established areas as opposed to the newly developed communities.

There is generally more disposable income in single family households in established areas as opposed to the newly developed communities. Newly established households will be spending most of their disposable income on furnishings, décor, renovations and/or repair. And lest you think otherwise, our industry is a luxury ticket item. Nobody NEEDS to study the martial arts. It is a choice. Hence it is discretionary spending on the part of the individual. It is suggested you target specific markets based on education level. Those with a college level education, a professional in their field, are more likely to be able to afford your services month after month as discretionary spending even if there is a downturn of the economy. People who live paycheck to paycheck will be hard pressed to make monthly payments in a timely manner. And when the economy downturns, and it will, these same people will drop out of your school in droves. They will either default outright or try giving you a personal sob story to get a break on their fees. You want to carefully consider whether or not you want to deal with that kind of grief. A small percentage of such individuals in your school may be okay, but keeping the percentage small is advisable for the long term.

As part of your thorough due diligence, also research websites like "Mover's" and "Monster's" (popular at the time of this writing). These websites can yield additional valuable information in your location selection process. Factoids on jobs available, salary level of those jobs and how many people are looking to move into a given area can be useful. In addition, sometimes statistics on crime trends can be found on such sites. This is something that most of your local chambers of commerce will tend to try and keep out of the limelight, and as a result, their literature. But such information is of tremendous strategic value to you in your search.

Demographic "overturn"

The dynamic of a given area (established vs. growing) is very important. You need a degree of overturn so as to not deplete your potential student pool too rapidly. Nobody wants to throw thousands of dollars and hours into growing their school only to find out that no new bodies are moving into the area. This means you will exhaust your "2%" rather quickly. Older, established communities may sound nice, but if all residents have been in that area a very long time and they tend not to move, this is not a good thing for the long term.

Financial status summary:

Single family homes = lower overturn = stability.

Condos/apartments = higher overturn = lower stability... but a higher number of new-comers means fresh market share.

Visibility vs. Ad Driven Traffic

Good visibility will pay dividends in rapid growth. Look for a location that has some sort of an anchor in the mall. Stores that cater to daily necessities are good anchors in that they draw a large cross section of the populace. A large chain grocery store is exceptionally good as an anchor in that all people from all walks of life must inevitably shop for groceries. This type of anchor will pay huge dividends over time in terms of awareness of your school in a given area. An Auto Parts Store may also be a good anchor for male shoppers. Just as a name brand clothing boutique may create a good draw for the female shopper. Thrift and or Dollar stores will tend to draw the bargain hunter. Specialized stores are exactly that, specialized, as is their customer base. Look for more generic anchors. You want anchors that will tend to draw the same market you are seeking. Ultimately, a good mix is desirable.

A fitness center may, or may not be, a good anchor for you. If they have no established classes and/or structure that teaches those things you may want to offer as part of your curriculum, you may be able to establish a sort of referral partnership with them. But, if they are indeed offering such items on their menu of services, you may want to stay away from that mall. Research carefully on anchors before you make a decision. It could be a costly one if you are not careful.

Good signage will far outweigh any sort of ad driven traffic. A good sign is more cost effective by far and is a permanent fixture in the neighborhood translating to overall awareness of your presence through time. Look for locations that are regulated, but not overly so in terms of signage. A nicely done, professional, large, lit sign is important. Stay away from malls that require cute but difficult to read characters in their signage bylaws. (see more about signage on; Page 12 of "Establishing A Marketing Plan")

Being Mall Anchored will equate to stability in your school because of the repeat clientele which will tend to be drawn to an established mall again and again for their daily shopping needs, maximizing your exposure.

Be advised that any amount of drive by traffic will do you very little good if it is moving too fast! Choose a location that affords traffic drive by at less than 45 mph if possible. Better still is a location that allows visibility from a major intersection so that drivers can see your sign while they sit waiting for their traffic light to change.

Isolated Strip Malls vs. Major Shopping Malls

Isolated strip malls are probably a good bet, all things considered, in most cities. They can afford a great deal of drive by and/or walk by visibility. They often have a major anchor or draw into the mall in the form of a large grocery store, etc. And they are often owned by individuals which are more receptive to start up business like martial arts. Oftentimes they are well managed by outside management companies which means that any issue you have will be addressed quickly and professionally. This becomes very important when your bathroom commode constantly overflows or your A/C units go down for example.

Large, consolidated, full malls are often not as receptive (think hefty bond) and square footage costs can be prohibitive to a successful business model. Management of large malls will tend to look at martial arts as "undesirable" because of the possible clientele that management perceives may frequent the school, potential parking problems and the perception of a "non established business model" which in their minds equals a likely failure. As a result, you will be hard pressed to change their mind unless you can indeed put up a cash bond to guarantee a year's worth of rent on the lease. Also, the square footage costs plus cam charges tend to be so very high that your initial 100 students tuition will go straight into their pockets instead of yours.

You may be tempted by an extremely Low Rent in a dead or dying mall (i.e. many vacancies) while you search for your location. But be very careful of such offerings. Keep up with your due diligence in such places. Ask questions of nearby businesses. Discover what was there before and why it did not survive. There may be a very good reason why that particular mall is so dead and you do not want to become one of the grave markers of failed businesses in its bone yard by not looking deeper than the cheap rent.

Likewise, you may be tempted by a seemingly great location that is not quite big enough and carries a hefty premium on lease. If the rent is so high that you have to count on part of the parking lot as an expanded classroom, it would be best to look elsewhere. You can run into problems this way. Complaints from neighbors may cause repercussions with your landlord. AND, you may not be covered with your liability insurance should any injury occur off of your immediate premises. Weigh such things carefully in the sense of the big picture.

Your Immediate Competition

The local players (other martial arts schools within a three to five mile area), large or small, are something you must familiarize yourself with. And you must do this in a complete manner. Find out what their operation is like, how they structure their curriculum, hours of operation, price points, what market they are after, all of it. And then keep current on changes they make. Do your reconnaissance on your immediate competition every six months or less. Adjust your strategies as you see fit. But keep your eye on the ball. Just because the school down the street is offering some ridiculously low promotion, don't you do the same. To start, or to participate in, any sort of price war would only potentially destroy both of you in the process. Instead, find a way to counter what they are doing by adjusting your marketing or sales strategies. Do not change the fundamental way you do business but seek to find ways to offer what you do in an even better way.

Be mindful of "oversaturation".

Even the "non professional" competitors will take a portion of your market share and you must acknowledge this fact as you look for space. These are the part time operators. These are the people who do not teach for a primary living but instead teach for a supplement to their main income or job, who teach just for the pleasure of teaching, or who teach because they have a simple passion about their respective arts. They will never have many students. But you must not ignore their presence in a given area. Even if each one of these types of schools takes ten students away from your enrollment per year, it can add up quickly. Remember, only 2% of the populace is interested. And these schools tend to take *and keep* their small piece of the market share via loyalty. Besides, one of these small time operators may come into a windfall of cash and suddenly decide to expand their operation. Do not underestimate the influence they may have on your local market over time. Factor them into your decision process as well.

Establishing a Marketing Plan

Many things go into a good marketing plan. But to be successful, you must truly adopt the perspective of *making a plan*. Marketing is always going to be a constantly evolving, sometimes trial and error, based entity in your school operations. This is a fact. Yet if you *plan*, as opposed to wing it, you can save yourself time, money and frustration.

First things first... Signage!!!

Your sign. So simple, right? You put a sign up and they will come. Only in the movies does this happen. Signage is one of the most commonly overlooked areas and a frequent source of many costly mistakes made by the new school owner/operator.

You need a sign that works for your business, your area and your environment. All must be considered carefully. Cost, yes cost is a factor. But cost must be outweighed by the others or it will cost you much more in the long run.

Some malls and/or buildings and/or cities will limit what kind of a sign you may have for the sake of conformity to surrounding buildings, zoning regulations, etc. But once you have all that figured in, there are a few very necessary considerations to make.

1) You MUST go with an internally lit sign of some kind. Forget about any other option. Any other option will not garner the attention needed over the long term.

2) Think of your signage space as the most precious of real estate, because it is and you must not waste even one inch of it under any circumstances. A balance of both your *message* and *easy to read* must be struck in design. Meaning blank/white space *is* a factor in design.

3) You MUST make the sign succinct enough that it may be read at the normal speed of traffic in your immediate vicinity. A good test is to drive the area yourself and see what types of signs catch and hold your eye. Color, size of lettering, background color, all play a role and should be carefully chosen before committing the money required for production.

4) You MUST NOT think specifics. Nobody knows or cares about what kind of family art or lineage you teach. They can't read such things flying by in their car anyways. Give them only the general information and give it to them AS LARGE as you can in your signage. In other words, KUNG FU will work much better than will CHOY LAY FUT KUNG FU and

it will be considerably cheaper for you to produce. You can and will tell them the specifics of what you offer once they contact you but it does not belong on your sign.

5) If space, zoning and money permit, a phone number inclusion is a very good thing. People who are whizzing by your sign may not remember exactly what the sign says when they go to look for it later. However, if you do consider a phone number, it must be a number that is EASY to remember. Otherwise, it is simply a waste of your signage space.

Now, On with the Rest.. Exclusivity Sells!

Have you ever wanted to join an exclusive club? Adults love the feeling of belonging to a place that is special. It validates the ego and one's sense of self. Oftentimes, just being a member is what is important to an individual. Whether or not they actually use the services to any great extent is often less important than the actual belonging to a perceived elite organization. Which then grants them the right to brag to their friends about the fact that they belong to such an exclusive club.

The Student "Selection" Process

Now truly, there will be few people that will wonder through your doors that you will actually want to reject. But there will inevitably be a few. It is suggested that you install an "Interview" process to promote the exclusivity perspective in the prospects mind. Do make this simple but effective. An effective tool is to employ a simple application for the prospect to fill out which is followed by a face to face interview with the head instructor who then "approves" the applicant.

Typically, we should make the applicant wait 24 hours before getting that "acceptance" call for psychological effect. But in some circumstances, where you feel the prospect in question is on the fence, you may wish to "approve" them that very day while they are still in your school and get them to do the paperwork on the spot.

Just be sure to remain completely neutral. Prospective students can smell "hungry". You do NOT want to ever send this message. The message should always be that it is a privilege to be a member of your school, and not a right. Your school is selective about who will be let into its hallowed halls for study and who will not be allowed inside.

The interview process, once in place, will allow you to exclude certain individuals or stylistic looks that can denigrate the market share you are trying to procure.

You should vigorously nix or veto all body “piercings”. They pose a serious hazard in a martial arts school as it only takes one incident where a body piercing is either struck or torn out on the practice floor for you to become liable for negligent behaviors. Any piercings of the face are especially dangerous. And they also send a negative message to the clientele that you most want in your school, meaning the affluent.

You should also curtail the visible tattoos for the same reason. The psychological message is not well received. A small tattoo, tastefully done, may be okay. But large covered areas can send the wrong message to the bulk of your student body.

Vigorously eradicate all the gang insignia you may see or come across. Even the potential of an association to any gang in your city will kill your school. Allowing even the hint of perception that you may have gang members in your school can drive other students away by the dozen.

Being mindful of all these mandates are especially true if you hope to garner a portion of one of the most commonly overlooked markets.. The female market.

The Female Adult Student – the often overlooked market

Many school owners, either by deliberate design or lack thereof, tend to discourage the female student from joining their respective school. Women are more typically found in Pilates, Yoga, or Aerobics studios because they are sometimes intimidated by the usually male dominate scenarios found in most schools which are, of course, run by men.

But this does not and should not be the case. You should construct class curriculums to be friendly to women with specific coursework made for them. Make room for them and their needs in your school, right along side of, but treated uniquely from, the male population in your school and you will achieve a dynamic that will feed itself in referrals over and over again. Try to think progressive, modern lingo in your marketing literature. Classes that emphasize “Self Empowerment”, “Centering & Balance”, “Health Regeneration”, etc. will all work well for such women.

Women have much to offer over the long term health of your school. Statistically, you will see a lower rate of default on your contracts with women. Women tend to be less likely to “blow off” their financial responsibilities than do men. Even if they are no longer attending classes.

Many women will enroll as a single (unmarried adult) and they usually come equipped with many friends, also single, that have many friends, who are single and or married, who have many friends, and I think you get the picture. Construct programs that draw the adult female student into your school. Construct an environment that is friendly to her and you will reap benefits of that mentality a thousand fold over the lifetime of your school. Cater to women with programs than meet their needs and you will see your bottom line grow exponentially.

Selling your program to a prospective member

So you've now decided what kind of student you want in your school, right? (If not, go back and read the section "The Student Selection Process" - Page 13). And now that you know what type of individual you're after, how do you go about getting them involved?

Everything starts with getting these people to your school of course. And there is much more on the process of getting them there elsewhere in this manual. Every aspect from the initial phone call or "cold call stop by" to presenting your school and its assets to them is covered. But for this section, once you have them there, what should you do to interest them?

Choose a booklet over a pamphlet for your in house distribution literature. Make it informative, but not overly large. Think four to eight pages.

Understand that almost everyone "skim reads" these days. Knowing that fact, lead their awareness by using bold, colorful, bullet points in your literature. Fill out the booklet with emphasis on key points of your school and the motto of your programs, your school mission statement and success stories of past or present students. Additionally, any impartial 3rd party studies on the value of what you teach. Medical studies, for example, can go a long way.

Never post or hand out a *hand-written* flyer, notice or signs to your existing students and/or prospective members. This sends a message of being unprofessional in your endeavor and will tend to cause the more discerning prospect (the one you want most in your school) to think to look elsewhere.

Make sure to check and re-check all handout literature for typographical errors. Misspelling a word also sends a negative message. You want your prospects and existing members to always know that you are attentive to every detail of their studies with you. You want them to always know that you are a professional.

During prospective phone calls, you must avoid at all costs phrases like; "Just come on down and bring your check book. You can start your classes right away!" This belongs in a used car lot salesmen's jargon, but not in that of a professional martial artist.

It is also recommended that you avoid the ever popular; "Try 4 Lessons for only $25.00." The professional adult student is too wise for this tactic. You are sending a message of selling your school on the cheap. The educated consumer or prospect knows that this technique is simply a structured sales gimmick utilizing private lessons designed for the sole purpose of getting them involved in your school. And that they will learn nothing in the process of four short lessons.

A better sales phrasing example would be: "Your initial lesson will be a personal orientation lesson which will prepare you to acclimate in our curriculum before attending your first group class." This shows that you are an experienced teacher, and care about the individual student's progress. This introductory lesson also gives you an opportunity to establish a bonded rapport with the new enrollee. There is nothing more comforting than having an "anchor" in the form of a familiar face when getting involved in a new place.

You can't satisfy Everyone

Understand that not ALL people will want to become a member in your school. A "hardcore" program will not appeal to the yuppie audience. Just like a "soft-core" program will not appeal to the muscle head mentality.

Identify and go after a specific audience. If you begin to receive a lot of inquiries outside of the audience you have selected as your market share, you can always consider adding a program for just that group. But make it SEPARATE from your main programs in all ways.

Diversification in programs is a good thing. Changing everything is not. Keep the programs you already have to keep the people who became members under those programs happy. Add things for new prospective members that conform to what you see in the new client desires. But you must always be mindful of the "greater good" of your existing student base. Such people are the lifeblood of your school and must never be forgotten, slighted or ignored. Change, is a delicate matter for all people. Institute change carefully and with much contemplation beforehand.

Forget about an "exit" interview

Over the years, we have heard many martial arts professionals advocate that you conduct a so called "exit interview" with departing members. Forget about this as it is a complete waste of time and resources. Departing members won't tell you the truth anyways and you can come off as desperate to keep them which works against your "exclusivity" ideology. And you want to project this attitude of exclustivity at all times over the life of your school.

A far more productive habit is to get such information on the "front end" of a student's tenure in your program. Try asking your incoming new members; "Why did you leave your former school?" Or, "Why did you choose our school over the other school?" The answers you get here will pay dividends yielding immediate information and possible perspectives which you can use to help you sign up new members and gain greater insights to your market share. You will also begin to build a catalog of data regarding mistakes made by other schools which you can use in the future as you shape and reshape your marketing strategies.

Advertising - Traditional Vs. Modern

Traditional wisdom says that the Yellow Pages are your best bet. Well, run one yes, but make it count. Size is less important than presentation. Make sure that the content of your ads is more information based as opposed to generic. Construct the contents of the ad to attract desirable professionals and beware of what the ad conveys to them. Any perceived deception will turn prospective students away. In example, any comments like; " Our one hour classes are actually 50-minutes" sends a negative message, making prospective students feel short-changed. This approach may work fine for a practicing psychologist but it has no place in a martial arts school. Your ad must reflect exactly what you do.

The disadvantage of yellow page ads is the commitment factor. Most books come out only once a year. And whatever you do, you're stuck for that entire year with that respective ad, both financially and for the sake of marketing. It either works or it doesn't, so make it count.

Modern Advertising, meaning the Internet, represents the most bang for you buck. At the time of this writing, Google controls over 70% of online advertising. You would be foolish to not take advantage of this fact. Online advertising is by far more flexible, targeted and manageable than is the yellow pages. You can adjust your respective ad on the fly, try different strategies and can even adjust the amount of spending as you go along. This flexibility has a very nice appeal indeed.

In addition to your online ad, construct an information based website. Make the website more about giving information in detail about your art, your programs and your school focus. And then subtly direct the awareness of the reader to contact you through discreet yet prominently placed email contacts and/or fill-in forms for the prospect to give you his or her email address, a phone number to be reached and a convenient time to call. Then make sure you follow up on such leads within 24 hours.

Do not construct a website that is nothing more than a glorified yellow pages advertisement. Do so, and you will alienate the web savvy prospect. This person is again the person you want most. All professionals now use the web for everyday business, shopping and recreational needs. You want to reach them in a way that gets their attention and by giving them pertinent information they seek about you and your school. Right up front on your website is the best way to do this. Force them to call you for such information and they won't. They will simply keep searching online for what they want. Their attention span can be very limited so you must capture and keep them there long enough to get a direct contact.

Consider employing "Viral Marketing" strategies as well. An example would be to construct several short "Youtube" video clips showcasing some of the programs at your school. Make them exciting and short, approximately 2 minutes in length. Think television commercial (get to the point and don't bore your audience with fluff) and remember to include a contact phone number, web URL and email address at the end of each clip.

You might also consider a blog or discussion forum. But if you do, you must update such things daily or they grow stale and can actually work against you. So be sure you are up to that task or don't do it at all. Stale content will turn people away. And, as a result of this tendency, either you or a trusted employee or student must monitor and contribute to the forum or blog daily.

Demonstrations (pros and cons)

Public demos can serve to initially help get your name out in your local city area. But if you are going to do them, be very selective about the choice of venue. Most people going to a demonstration are not at that venue for that purpose. Your demonstration is solely entertainment value for them.

Good demonstrations do take a great deal of your time in preparation and often for very little direct reward. If you do demonstrations at fairs, mall openings, car show events, etc., expect that you may not sign even one student as a result of your time and efforts. But, it does allow people in your area to know that you are now there. Make sure to hand out flyers with a business card attached to all people that attend. This is low probability blanket marketing. But if you have the time, it doesn't hurt.

If at all possible, try holding your demonstrations immediately in front of your school and offer an incentive package for attending prospects. This will be far more productive as the attending people are there for that purpose. A good plan is an annual demo for Chinese New Year. You should advertise it well beforehand allowing at least two months prior. Be prepared and ready to hand out your incentive packages during the demo at an opportune time. Having students do this is best. Choose enthusiastic, people oriented students for this task.

Coupons, Flyers and Door Hangers

Of all the above devices, the most useful is probably the flyer if distributed via a stack in your local martial arts supply store. Or, in the prospect that you can convince your local Gym that you are not in competition with them but rather in compliment to them, a stack in their facility (probably in exchange for one of their lead boxes in yours) may work well.

Door hangers are again blanket marketing. A lot of time for very little reward. Good for start up perhaps as it is relatively cheap to execute.

Coupons seldom work as you will once again be selling your school on the cheap and those that are looking for cheap will not likely stay long term. So much time is spent and taken away from your loyal and existing members to service these short term "thought I'd try it's".

Free Classes = 0 dollars in your pocket. People looking for free are not serious. This counteracts your model of exclusivity. Do not do free, ever.

Lead Boxes can work in the right businesses. But if you use them, make sure you service them weekly for restocking and/or recovering the actual leads. Twice weekly would be better. A lead box immediately outside your school door for prospects that may stop by during off hours is a very good idea. Stock it with something that is not too expensive, like a single page folded handout. Keep it simple.

Targeted Mail

This is one of your most expensive advertising mediums next to radio and/or television advertising. Targeted mail should be used sparingly and must be timed appropriately. There must be an incentive to make the prospect call. A "New Year's Resolution" package for example. Otherwise, the financial drain can be quite drastic in such campaigns. Non specific or generic direct mail is a complete waste of your advertising dollars. Mailing a non specific ad is tantamount to serving steak to a room full of vegetarians. They did not ask for it and will not eat the meal. Structure your ads in a specific way, prompting action. Personal initiative is a necessary ingredient for a person wanting to become your student.

Be Professional!!!

In all devices, MAKE SURE that they are done professionally! Nothing hand written. All must be spell checked, grammar checked and proofed for mistakes! Prospective adult students that you wish to attract and keep in your school are well educated professionals themselves. Nothing turns them away faster than an unprofessional marketing campaign.

Anecdotally, we once dismissed a person out of hand who had approached us with a potential business deal because he had used misspelled two words in his letter to us. He had spelled out "Quailafactions" instead of qualifications and used "Intrested" in place of interested. Our immediate perspective was if he were not careful and meticulous enough to spell check his letter to us, we certainly did not want him in a position where we would need to rely on him.

Business/Marketing Plan Sample

You want to draft a business/marketing plan for two reasons. One, you will need this for "Lease Holder Satisfaction" when you find a building or mall you want to lease for your school. And two, and no less important, it will keep you focused on the tasks at hand for starting your school. Make it as detailed as possible.

An outline form is fine but a detailed outline just the same. Include the following:

Projected start up costs (construction modifications of the shell, signage costs, equipment).

Projected initial advertising costs (internet, yellow pages, flyers, radio, T.V., coupons).

Projected sign ups (from your demographic research and spot reconnaissance of your chosen location, daily walk by traffic).

Projected time of "break even" in your business plan.

All these things will serve a dual purpose of making your leaseholder more confident in giving you the lease in the first place *AND* keeping you on track with your business!

Need More Help with your Marketing?

MASTER MIKE PATTERSON:

He is a *Highly Successful Owner/Operator* for Adult's Only Chinese Martial Arts Schools. He has acted as *Publisher* for an International Martial Arts Magazine. He has been and still is a *Producer* of a full line of Internationally recognized Instructional Videos since 1988. He was one of the first to utilize the world wide web for promotion of the martial arts and he continues to be cutting edge in all aspects of martial arts promotion. He is the *Founder* of the Innovative Martial Program R.S.P.C.T. (realistically structured progressive combat training). He has 44 years of Martial Arts experience at the time of this writing.

MASTER ANDREA CHENG:

She was a *Million Dollar Manager* in a leading Fortune 500 Company. So effective that she became one of the top *Certified Training Managers* of that same Fortune 500 Company. She was consistently a *Top 2% Performer* at the national level of a leading Fortune 500 Company and a *Fiscal Performance Troubleshooter* for that same Fortune 500 Company. She has been and still is active as a *Consultant* for personal training programs tailored to the business professional. She has 29 years of Martial Arts experience at the time of this writing.

Hire us as outside consultants to troubleshoot your operation. We can help you grow.

email us through: www.hsing-i.com

Negotiating a Lease

Once you find a building or two that you are interested in leasing, you will need to contact the landlord or leasing agency. Make sure you have your business/marketing plan done as you will need to both refer to and probably submit a copy to the person in charge of leasing the building.

Understand that in lease negotiations, what is said upfront is often negotiable to a degree. And if you play your cards right, you may be able to get far more than you thought possible from the landlord and/or leasing agency.

Above all else, remember that in any negotiation your position is strengthened if you are willing to walk away at any time from the deal. If the agent senses you are "hungry" you will end up on the short end of any negotiation and/or concession.

Tenant Improvements

"TI's" or tenant improvements can often be negotiated as part of a lease deal provided you are being reasonable when asking for such improvements. A landlord or leasing agency is not at all likely to be willing to foot the entire bill for a second bathroom, but they may give other concessions if you really feel you need to build one for your business. They will almost certainly be willing to finish out the shell of the suite, build some additional partition walls, perhaps put in some reasonably inexpensive flooring and the like. But each landlord is different and will only go so far in granting such requests. Much will depend on how long the space has been open and how the current leasing market has been flowing. If space is plentiful and times are slow, you will be far more likely to get most of your concessions in the form of TI"s than if space is hard to find and the leasing market is healthy.

Free Rent

Free rent, sometimes as much as six months depending on the market, can often be negotiated as part of the lease and you should not be afraid to ask for this consideration. Your negotiation position on this issue will be strengthened if you have done your homework in your business plan and are able to show the agent a projection of when you expect to break even and have a sustainable cash flow in your school.

The leasing agent may then come back with a counter offer such as one year at half rent, which is *still* negotiable. Stand firm on what you want and back up your desire with facts, figures and projections from your demographic research. The leasing agent wants a tenant that seems to be a "good risk" and your ability to show such organization within your research will go a long way in sending that message. This will in turn get you what you want most in your negotiation efforts. Stick to the facts and continue to prove your position with hard data derived from sound research on your venture.

Cam Charges - Hidden charges that you simply MUST be aware of...

Most of the time, when you inquire as to the square footage costs on a lease of a building or space of interest, the leasing agent will not include what are called "cam charges" or "triple net" charges in that quote. These are charges that the landlord will wish to collect, in addition to the cost per square foot, on any lease negotiated. They are used for things such as structural repair to the building, holiday décor, building renovation or remodeling, etc. And they can add up to a substantial increase in your overhead.

Find out when the last renovations were done on the building or mall. This will keep you safe from any really major unexpected "gotchas" in terms of your monthly cash flow during your lease.

You should look to include a clause in your new lease agreement that will cap your cam charges increase on a yearly basis. If the landlord or leasing agent is not willing to cap the cam charges reasonably within your lease term, for say cost of living increase per annum, look elsewhere. The leasing agent may be trying to hide some sort of looming major repair and stick you with a portion of that cost.

Triple Net/Cam Charges Table (monthly):

Sq. Ft./Cam	10 cents	20 cents	30 cents
1000 Sq. Ft.	100.00	200.00	300.00
2000 Sq. Ft.	200.00	400.00	600.00
3000 Sq. Ft.	300.00	600.00	900.00

Anecdotally, we were once in negotiations for an expansion school into the Encinitas Area in San Diego's North County Region. We found what we thought was the perfect location, situated on a busy intersection, right on the corner next to a Block Buster Video Store. The other mall anchors were a nice mix, including several clothing stores, a large grocery chain store and several food anchors for the surrounding population. Overall, a very good find. Negotiations were proceeding very smoothly. The leasing agency seemed very eager to accommodate our every desire. We were granted nine months at half rent, plus two months up front free rent for setup of the facility. They granted several tenant improvements in the form of partition walls, etc. But when it came down to the final once over by our attorneys in the lease itself, there was no cap on the cam charges. When we asked about that, the leasing agent was unwilling to put such language into the lease, which in turn made us suspicious. So we investigated. And through our due diligence discovered that the landlord was planning a major renovation to the property in the coming New Year, which we would have been saddled in part with for costs to the tune of several thousand dollars. This extra expense would of hit us within the first six months, making it very difficult to break even on that property within the first year. So, we walked away from the deal.

If this is your first time negotiating, hire a commercial realtor to be present during all phases so that you may learn. Next time, you may not need him/her. DO NOT make the mistake of going it alone the first time. It can be a very costly business lesson.

Walk Through Before Signing

Do a thorough "walk through" on all changes made to the premises before signing your lease. The time taken in correcting electrical, A/C, Plumbing, etc. may cause delays in your grand opening if you are not proactive in taking care of it now. And you want to make sure that all things have been completed satisfactorily to signing. Once you sign, you lose some of your "power" of persuasion over the landlord or leasing agency. A signed contract means you are in and they may now treat you accordingly. An unsigned contract means they must still make you happy in every way or risk losing you in the deal. Also, setting the tone early on that you will not settle for less than what was promised you is always a good thing for further relations down the line with the other party.

See the next page for an essentials walk through checklist...

"Essentials Walk through Checklist"

1) Turn on all air conditioning and/or central heating units on the premises and let them run a couple cycles to make sure they are operating properly.
2) Bring a small electrical device to check all electrical outlets and check each and every one for power.
3) Turn on all faucets. Check for both cold AND hot water.
4) Flush all toilets on the premises and watch inside the bowl as they fill.
5) Open and close all windows and doors on the premises.
6) Do a general visual "once over" for anything that just doesn't "look right".

BTW, "Stick to What You're Good At"

Don't think that you are qualified to look over your lease yourself and handle the language without professional help. Spend the money for a legal consultation on such issues. And then have a qualified attorney go over the language with a fine toothed comb looking out for your best interests.

DO NOT make the mistake of feeling you do not want to, or do not have to, spend the money for such services. You DO need to have professionals help you in such circumstances as leasing language can be tricky at best and outright misleading at worst.

An excellent example of this perspective can be found in the realm of the all important *"Liability Insurance"*. This is a must have of course! And you will need to supply the proof of having liability insurance during your lease negotiations or no landlord will touch you. BUT, and this is important to understand, liability requirements will vary from state to state and from landlord to landlord. Only an attorney can really advise you properly on how much you really need to carry and balance that with both what your state requires, and what your landlord may desire for their own protection. You are not equipped to make such a critical decision without expert advice.

See The Next Chapter for more on this perspective.

Stick to What You are Good At

Knowing one's own limitations is something we all preach philosophically to our students. And we also encounter our own limitations as we grow within the arts. Since the martial arts teach to overcome such limitations through training, we tend to develop such a mindset in relation to other parts of our lives. This is only natural. But just as we have become experts in our own fields, there are others that have put on their respective "Gi" to acquire the same level of expertise in their fields of endeavor. We should respect that we cannot know all and humbly ask the benefit of these same people when the circumstance warrants.

Pay a Professional

As practicing martial artists, we get used to problem solving. We deal with it on all levels. In the beginning of our training, as students, with regards to correcting form and/or modifying a technique making it more viable for us in actual use. Later still, as teachers or instructors, we hone our problem solving skills further in the form of motivating students, helping them to overcome their own weaknesses. As a result, we can begin to develop a great deal of confidence in our own abilities and this is a very good thing in the martial arts world. But, and this is a big but, these same skills do not necessarily translate to other areas. Making such a mistake in thought can cost you dearly. You are only truly as knowledgeable as your own immediate and direct experiences have prepared you to be.

You are a martial arts professional. Your parameters all fall within that arena. You are skilled at your respective style and its method, techniques and trainings. You are skilled at imparting that same knowledge and understanding to others. And while it is true that you may also learn and acquire much additional knowledge over years of being in business, you must still understand the limits of such additional acquired knowledge. Not understanding this can cost you much grief not to mention monetary loss if you overstep your boundaries with false confidence.

Legal Forms

Anything regarding the legalities of your business should always be placed under the domain of a qualified attorney. Legal Forms such as a waiver of liability, which is an absolute MUST in your business, need to be placed into the hands of a qualified professional.

Laws vary from state to state and an attorney is not a luxury but a necessity when it comes to protecting you and your business. They will be able to properly advise you on what you can and cannot do in regards to insulating yourself from any potential lawsuit that may arise from your teaching activities.

Make the mistake of trying to shortcut this expense and it could cost you all that you have worked so hard to build. One successful lawsuit can completely drain all of your cash reserves, effectively wiping out your school. Even an unsuccessful suit can cost you thousands to defend. You want to make your legal documents as airtight as is possible in your state to discourage such events from coming to fruition.

Additionally, different levels of waiver may be required, depending on what you offer in your curriculum and this can vary from state to state. For example; If you offer non contact sparring in your school verses full contact sparring, both the insurance companies AND your respective court systems may look upon these activities very differently. It is very wise to find out before it becomes an issue.

Contracts

Contracts are an absolute must in a school that wants to show and maintain a respectable profit. Once you decide to utilize a contract, you will need an attorney to structure the language based on the laws of your state. The law will, at a minimum, require what is called "consideration" for both parties.

This means that the contract in question cannot be written in a one sided manner to which only one of the parties has more rights within the contract than does the other party. You will need an attorney to properly balance this equation and include the appropriate protective measures for your interests at the same time.

Different contracts for different types of programs may also be required. A contract written for group study will not be exactly the same as a contract written for private study, or for a contract written to include both group and private study. You may need a slightly different contract for each of the programs you offer. And again, an attorney is the only person qualified to tell you what is necessary in your state.

Assume nothing and ask everything when consulting your attorney.

Lease Approval

For a lease approval, it would be prudent to have both your attorney AND an independent real estate leasing agency or agent look at the language in the Lease. Unless your chosen attorney lists commercial real estate as a specialty in his or her professional credentials, an attorney alone may miss some things in the leasing language that a real estate professional will see.

Commercial real estate agencies, like all industries, have their own way of looking at things and have their own unique industry terms and language. Besides, when it comes to locking yourself in for a five year stretch on a lease, two views from two professional heads are definitely better than only one. Five years, or even three years, is a very long time and much can happen or change in that span of time. More information can only assist you, not hinder you.

Again, assume nothing and ask EVERYTHING if this is your first time through. This mindset will pay major dividends not only in this deal but for any future deals as well. Asking questions now will help you to understand, and process, for the future.

Setting up your Facility

I know, I know. You get a room, hang a bag or two and you're set, right? Wrong! *If done right, your school layout can be your best sales tool, but if done wrong it can really work against you.*

Storefront

In the early days of martial arts schools, a closed school front was the norm. To not allow any people to look or see inside was what everyone did. Some still do this and certainly if you are in a location that allows those that pass by to press their noses to the window, you may want to

subtly discourage this behavior. But there are things you want to do and things you definitely do NOT want to do in regards to your storefront. Your storefront is like your handshake. It is the first impression a prospective student receives about your operation. It needs to be as positive and welcoming as possible.

Open Windows

And no we don't mean for fresh air. Although that is important as well. You want a relatively open window so that there will be drive by visibility of your class in action. The ability for the prospective student to see an abundance of activity in your school as they drive by can be a huge draw to their awareness. The caveat here of course is that an empty school, devoid of any activity, can have quite the opposite effect. So in the beginning stages of start up, when your school has few students, you may want to consider a slightly different strategy.

Oriental type screens can provide a nice partial and attractive covering for your windows and, in the beginning, if you want more closure when students are sparse they can be backed with painted wooden panels to obscure view. It's very cost effective and nice looking. Then, when your student body fills in a bit and you want to maximize the wow factor, the wooden backings can be removed leaving just the oriental screens. A driver coming by the premises will see activity through the screens which will peak interest in your school.

So remember, once you have a decent student body, avoid full closure of your windows. Let any and all the passersby see the activity of bodies doing form, weapons, etc. within your school and remember to have literature near the door ready to hand to them when they stop.

TV on a timer

You may consider installing a reasonably sized television or monitor to run video presentations during hours when your school is officially closed. If you do this, make the presentation action filled. Not one minute of the presentation can be slow paced or boring. You don't know how long a person may stand and watch your presentation. You want whatever they do see to stay in their minds once they walk away. The history of your art may be important, but it does not belong here. The presentation should reflect what you do and offer by showing exactly what you do and offer. Keep it upbeat, positive and NEUTRAL. Meaning it is best not to show free fighting as it may send the wrong message to the casual passerby. Focus the video on the positive aspects and not the violent. Showing technique is fine and actually desired, but make it about *self defense* not *fighting*.

Lead Box

This is an absolute must! The ideal lead box is place outside your school in a prominently visible place. For example, right next to the main entrance. A great lead box has two attributes. It has a contained, covered area for literature that a prospect can take with them. AND a closed and locked slotted box where a prospect can leave their name and a call back number if they so desire.

The literature to be distributed should be brief, probably only one folded page of cardstock. Cardstock is so much more tangible and has the appeal of a giant business card. Whereas paper feels more like a flyer and is more easily crumpled up and discarded. Do not put too much information on this handout. Keep it simple. Your school name and logo, address, phone number, web URL and hours of operation are primary. A few bullet points about your school programs and its best features and a colored graphic or two. You want it clean and easy to read, not overwhelming.

The slot that allows a prospect to drop in their name and contact number needs to be well marked. And you must supply the proper paper, well organized, and a pen attached to the box. Whatever way you organize this lead box, it must be well maintained and neat in appearance. Remember, it is part of your storefront handshake. Maintain it meticulously and at all costs.

Trophies - The Gleaming Draw

There was a point in time that all schools liked putting trophies in their windows to showcase how effective they were at training their trophy winning students. WRONG PERSPECTIVE! Keep the trophies in a designated area INSIDE your facility. It is nice to let people know that you are able to train such prize winning students. But, and this may be news to you, most adult students do not care about winning trophies. They care about how the art you teach can enrich their lives. And to most adults, *trophies in the front window = Pawnshop*, when driving or walking by your school.

Your front window space can be, and should be, used in far more valuable ways. You are not selling trophies. You are selling martial arts. Keep the store front in line with what you actually sell and don't confuse the issue for the casual passerby.

Internal Presentation

Many school owners even today seem to think that function outweighs all else when it comes to outfitting a school. This could not be further from the truth. Function is certainly a concern, but that is not the only concern in constructing your school's internal layout.

For the adult market, you are selling more than just martial arts classes. You are selling them an entire experience. They want all the trappings that go along with the mental picture they hold in their heads of the "venerable Kung Fu Master" and the school in which he or she resides. You want to cater to that vision and give them exactly what they want. Make your facility echo that mindset in every possible way.

Be professional, not corny. Be functional but not "gym like". Give them a Kung Fu Temple of a sort and they will love walking through the door again and again.

Ambience Sells

The adult student wants to see a tastefully decorated, traditional *AND* functional school. Using traditional Chinese wall art, not too much to be tacky, but just enough to add ambience is both appropriate and desired. A small statue of Kwan Kung, the patron saint of martial arts, is equally appropriate perhaps on your desk where you make final sales and can also serve as a conversation piece useful for "bridging" with your new student a bit while they are filling out the sign up paperwork.

Stay away from anything resembling an altar as it can send the wrong message. You are not selling religion. If you do teach philosophy, work it in as part of the overall package. It is a big mistake to overemphasize this in your programs and worse to blatantly showcase it in any of your décor. YOU may be Buddhist and proud of that fact. But many of your students are not and do not wish to be so.

Equipment Sells

Line the walls with various equipment. A well equipped school spells professionalism and functionality. Traditional Chinese Weapons, *neatly* displayed, training apparatus such as heavy bags, rhythm bags, striking dummies, tumbling mats, etc. *All must be neatly organized!* It does not matter whether or not all of your students use such equipment or if any of them do for that matter really. The fact that such equipment is there for them spells value to the adult student. Nothing spells authenticity like a shining display of traditional weapons combined with apparatus that will allow development of skillful execution of technique. This is marketing. Don't cut expenses here.

Amenities Sell

Think Student Library for example. A library can be placed in a small corner space. It is low cost and adds tremendous perceived value for the adult student. We have done this successfully in an 8 by 10 space, walled off with a door for a sort of private "hang out" for students before or after class. The collection of books eventually topped 500 and we did it all through student donation. Our only costs were shelves, a few chairs, some acupuncture charts and artwork for the walls. It never failed to make a strong impact on any prospective student.

Also, if you have the space and inclination, a small "natural juice" dispenser/machine can also be a good idea. You won't make much money on it but it will create a nice feeling in the student body. If that is too much for you, a hot tea dispenser also works nicely.

Cleanliness Sells

Adult professionals will NOT come to a dirty school. They want a clean facility to train within. *Clean = Professionalism*. It shows people you care about their training environment which in turn translates directly to increased sales and long term loyalty in your student base. Put this into your mind and keep it there… *Clean = Professional = Sales*. Never, ever forget this and maintain it always both inside and outside your school environment. Also, check and maintain bathrooms regularly. Police outside your storefront regularly, several times daily is best. Keep all equipment clean and well maintained. Replace old equipment when it begins to look tired. Keep all shelves, pictures, artwork, etc. dusted regularly. Think "white glove" and then execute that everywhere in your school.

For all announcements, designate and hang an official "Bulletin Board" This will allow you to present a clean facility by preventing random posts or flyers on your walls. Random posts on your walls will appear cluttered. A proper bulletin board will direct the focus of your student body, representing your school with well organized professionalism. Direct all new students to check the bulletin board daily when coming to class for anything pertaining to them, such as upcoming seminars, etc. Keep it current. Do not let outdated announcements remain even one day beyond their expiration.

The "Walkthrough"

Now that you have done all this work to present a clean, organized and professional school loaded with amenities you must work into your sales pitch a "walkthrough" of your remarkable facility. If you do this right, the facility will practically sell itself.

The person doing this must be trained to hit all the highlights of the school, talking about each of the amenities provided for the student and what each one means to them. Rehearse this a few times before going live with a prospect. Speak with passion for your school as you conduct this tour and such passion will invade the awareness of the prospect. Be succinct but thorough in your presentation. Do not ramble and do not lose focus. A rehearsed "script" is best but with a little variance so that it does not *sound* like a rehearsed script or "canned" to the prospect.

Starting a school on a tight budget

We started our first commercial school on $1500 in 1981 in a one room facility. Over the next ten years we grew it into a multi hundred student base and a four room facility, complete with separate restrooms and a student library housed in another small room.

It CAN be done. But you must make some concessions and be willing to do most, if not all, the work yourself in the initial phases of startup.

We did this startup in an industrial space. This space was situated on an access road which was parallel to a major thoroughfare. The visibility was not great. There were no major anchors.

But, the landlord was receptive to signage locations that improved our visibility to the drive by traffic. This, plus the yellow pages ads, allowed us to survive, thrive and grow. This is not what we would recommend by any means. It's a hard way to go in the beginning. But if you must, you can try with an eye to relocation when you develop enough of a student base and enough cash reserves to make such a move. Try not to get locked in to a long term lease as you may want to move in just a few years.

Budget initial construction costs from the perspective of making the space both appealing and functional. Plan this phase of construction with an eye thinking to "upgrade" it easily when you have established a bit of cash flow. For example, think: painted floor vs. carpeting vs. mats in that progression. Framed "posters" vs. actual art on the walls later on. An office "space" vs. an actual office with walls later on. You get the idea. Plan and execute your upgrades in a timely manner to continue to lift the appeal of your space over time.

Another route to take is start up in a shared space. Dance studios are often struggling in their space to make ends meet and will be open to sharing their respective "down times" with you. However, this will severely limit your growth potential and as such you should again be only doing so with an eye to move as soon as you possibly can to a more suitable locale that is yours and yours alone.

A space of your own is critical to evolving a complete curricular structure in terms of allocation to peak and non peak hours depending on your market. Being confined about when you can and cannot offer specific classes is just not in your best interests for long term growth.

Constructing your Curriculum

A good curriculum needs to take into account three things:

(1) It must be engaging to the student. They must always feel enjoyment with involvement.

(2) It must systematically progress them through the system which they are there to study. They must walk away from each class feeling that they have learned and retained something of value to forward them in their respective studies.

(3) It must be profitable to the school. The structure must allow for long term retention of each student which translates to profitability for the school.

Small Bites w. Rewards

Students are goal driven. Your curriculum should be structured to capitalize on that mentality. For the adult student, you want the individual segments of their studies to be challenging, but not so much so or take so long that they begin to feel like they are "not getting there" and, as a consequence, begin to lose focus.

Your goal is to help them stay self motivated and enthusiastic about their own progress within the art. An adult does not want to be coddled but they also do not want their spirit broken. They need stimulation in their studies on a regular basis to prevent feelings of boredom or that they have hit a wall in their study.

Rotational Blocks - Lesson Plan

An easily implemented and effective tool for adding stimulation and diversity to your programs in general is to extract key common elements from your respective style and then to structure rotational "block study" lesson plans.

Systemization promotes confidence in the student base and adds perception of value overall. In our school, we teach three Chinese Art forms. And although they all have unique flavors and different forms, concepts, methods, etc., there are some overlapping general ideas. Extracting such overlapping elements and then systemizing those same elements into "blocks" that you can rotate in and out of your ongoing curriculum creates a stable structure that students can readily avail themselves of which in turn creates perception of value.

Elements such as throwing, locking skills, kicking techniques, hand combinations and ground skills are a few examples. Look at the respective arts you teach from a component perspective and then determine suitable rotational blocks. Another direction to take is to structure blocks that exactly and specifically parallel your ranking structure.

Adult/Professional Friendly Programs

Another way to increase perception of value in your school curriculum is to create programs that are geared to specific portions of your student body and/or those you wish to attract.

As our school evolved over the years, we began to notice a greater influx of adult professionals attending our classes. These people make great students as they are stable and reliable in terms of finance and have a genuine interest in all that martial arts has to offer them.

Yet the one thing they *do not* want is to get injured at class so that they are prevented from showing up to their jobs the next day. They may want to learn about the combative aspects, but not at the price of black eyes, cuts and abrasions and injured joints which may hinder their day to day work performance and/or image.

As a result, we created and trademarked the R.S.P.C.T. program. (standing for: Realistically Structured Progressive Combat Training). This is a program that focuses on bringing the adult professional slowly and methodically into true self defense and combat training. It employs a systematic approach to such training by introducing layered modules, delivered at a pace which they control, minimizing any injury in the process. This program allows such an individual to pursue the martial aspects of what they are studying without the fear associated with normal "free sparring" activities. Too many schools just simply throw all students into free sparring without adequate preparation, which leads to injury, which leads to fear of involvement in such activities. The adult professional will appreciate a less haphazard approach to such things.

Again, professionals do not want to get bruised, hurt or sprained during training as this would affect their professional image. They simply *cannot* show up to a client meeting the next day with a black-eye, bruises or gashes on the face, hands, arms or worse a broken bone. You MUST be empathetic to this fact.

Keep Pace with a Changing Sales Environment

A good school owner will keep his or herself abreast with the current trends and shifting public awareness and then offer classes accordingly. We are not advocating that you jump on the band wagon of every fad and trend that comes your way. Quite the opposite. Merely be alert to the changes in your potential client base. Pay attention to the types of individuals that are coming to your school seeking instruction and make sure you have a program to offer them.

In example: When we opened our doors on our first school in 1981, the average student profile walking through our door was a 20 something male, full of energy, bravado and testosterone. As a result, the entire curriculum was structured around high energy, workout related classes and classes that dealt with combative aspects using two person drills, pad/bag work and free fighting activities. Very little emphasis was given to the softer aspects of meditation and chi kung (qigong) or the more intellectual pursuits like Weapons theory in those days.

But with the introduction of MMA (mixed martial arts) into the mainstream of the industry. We began to notice a gradual but continual shift in the average student profile. Many of the 20 something's were now pursuing MMA first and foremost due to the media attention it had been receiving. Whereas, older males were now interested in something that would give their lives discipline and focus. Also, as Tai Chi (Taiji) became more known in the West, more and more middle aged women began wanting to pursue those softer aspects of the study.

As a result, we began shifting the curriculum slowly and steadily to accommodate this change. And by the mid 90's had a structure that was about 60% softer study and intellectually oriented coursework and only 40% coursework structured toward high energy and combative aspects.

You must be both alert and agile in your adaptations to such changes in your market or you risk losing a substantial part of your revenues or even your school as a result. Sticking to something just because "that's the way we've always done it" is both foolish and counterproductive to long term sustainable growth of your school.

A Systems Driven Curriculum

All too often, a martial arts school will cultivate loyalty to a specific instructor. It is far better to cultivate that same loyalty toward your overall school. You must choose style and protocol over personality to achieve success. This will allow greater flexibility should you wish to introduce change in your curriculum. And if one of your developing instructors encounters a life changing circumstance, it will be easier to edge them out of your structure more seamlessly. All assistant personnel should be cross trained in all aspects of your class curriculum. There should never be any one individual who is so specialized that they cannot be easily replaced. The only exception to this is the actual head of the school.

Cultivate systems and protocols for everything in your operation from taking calls, interviewing prospects, giving tours, and teaching specific classes. Script everything out and then teach all personnel to follow the protocols specifically without any deviation. A systems driven school runs smoother in all aspects of operations which translates to higher revenues. For example: You would not want someone who has not been properly trained in phone protocols to answer your phone. It is easy to understand that the result may very well cost you a good prospect.

Ranking Structures

Too often, especially in Chinese disciplines, ranks are too few and far between. Adults need short term goals to keep themselves motivated. Too much time elapsed between rewards for their efforts can and will translate to feelings of spinning their metaphorical wheels, which can in turn cause them to dropout.

Structure your ranking so that they can achieve such small rewards regularly. You don't want too many ranks as this may seem overly patronizing to an adult. You want just enough bars where they can jump over one of them every few months. Make sure the criteria is clearly laid out ahead of time and also make sure that at least some of your classes have been specifically designed to help them achieve those same goals.

As example; one of the arts we teach is called Hsing-I Chuan (Xingyiquan). Our respective style of the art contains 28 unique hand sets, not including weapons. We evolved a curriculum that contained 6 main ranks from beginner to 1st degree black belt (sash for us). And then had sub-ranks within each main rank, of three tiers each denoted by chevrons, on the associated main rank sash color. This gave us a total of 19 steps to 1st Degree Black and allowed us to break the curriculum into much more digestible bites for the student body. This meant that a reasonably focused student could advance to the next tier every couple of months or less, which in turn kept them self motivated in their progress.

Our curriculum was composed of classes that taught all the required forms, other classes that taught all the required combative knowledge and still other classes that taught all of the required theoretical and intellectual aspects of each tier of study. Again, a systems driven school will translate to efficiency and a real perception of value in the mind of the adult student.

"Lateral" Study - Diversity Sells

There is no question that diversity sells. The more you can offer a prospective student, even if only from their point of view, the greater the perception of value. You do not have to be able to teach more than one art form to accomplish this. You simply need to "compartmentalize" the way you teach the one art form.

Let's say you teach one art. Of course it will contain some way of self practice (form, kata, hyung, sets) so this becomes one type of class. If it contains weapons, this becomes a separate class. If it contains throwing, still another. Locking skills are still another separate class, etc., etc. You are now selling a diverse curriculum in the prospects mind offering many directions of study, as opposed to offering just "one thing".

You will also find that breaking your discipline up like this will force you to make lesson plans for each component and this will translate into more thorough coverage of each aspect of that art form which will in turn produce more knowledgeable and skillful students. And, after all, is that not what we all want as teachers?

In our school, we taught three different disciplines which allowed tremendous "lateral" study within our curriculum. The schedule itself fairly sold the school yielding very strong perception of value in the prospects mind; "You mean I can study *ALL THIS* for *ONE FEE*?" YES!

The 40-Hour Workweek

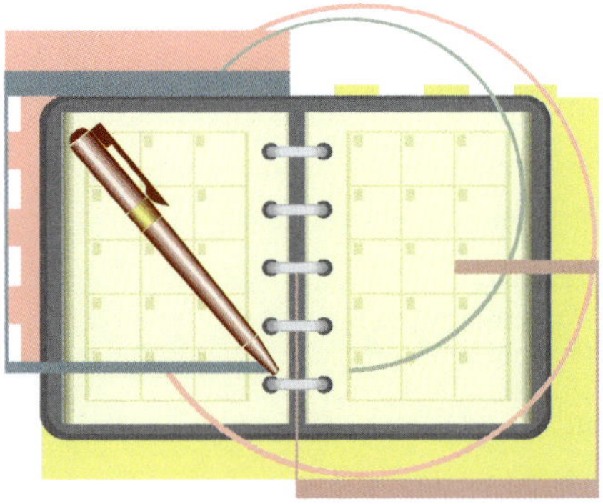

With our enthusiasm to build our schools up to the potential financial powerhouses they can become, we can sometimes lose sight of ourselves. And trust me, we learned this the hard way and you do *not* want to make that mistake. Now there may be many periods when you must and need to put in the extra time. During startup phase is a good example of this. But be disciplined about it just as you have always been disciplined about your own practice. You must take steps to preserve your own longevity. After all, you ARE the product you are selling! And your body must be cared for right along side any other desire, need, want or necessity.

A 40-Hour Workweek allows you to recharge. You must allow time for your own self-practice and health regimen to prevent "Burn Out" in yourself. This is very important for you and your school's longevity *AND* it is achievable within the context of your goals.

Potential Example of Class Schedule

	Monday	Tuesday	Wednesday	Thursday	Friday
12:00 p.m.	Private Lessons	Private Lessons	Private Lessons	Private Lessons	Private Lessons
12:30 p.m.	Private Lessons	Private Lessons	Private Lessons	Private Lessons	Private Lessons
1:00 p.m.	Private Lessons	Private Lessons	Private Lessons	Private lessons	Private Lessons
1:30 p.m.	Private Lessons	Private Lessons	Private Lessons	Private Lessons	Private Lessons
2:00 p.m.	Private Lessons	Private Lessons	Private Lessons	Private Lessons	Private Lessons
2:30 p.m.	Private Lessons	Private Lessons	Private Lessons	Private Lessons	Private Lessons
3:00 p.m.	Private Lessons	Private Lessons	Private Lessons	Private Lessons	Private Lessons
6:00 p.m.	Beginners Intermediate	Intermediate Advanced	Beginners Intermediate	Intermediate Advanced	Beginners Intermediate
7:00 p.m.	Intermediate Advanced	Beginners Intermediate	Intermediate Advanced	Beginners Intermediate	Intermediate Advanced
8:00 p.m.	Beginners Intermediate	Beginners Advanced	Beginners Advanced	Beginners Advanced	Beginners Advanced

Group Classes - It is possible to Schedule three one-hour classes per weekday evening Monday thru Friday (in this example 6-7; 7-8 & 8-9 pm respectively) at two classes per hour with two instructors if the facility permits (meaning a two classroom facility). You must first determine the peak days and non-peak days with respect to your own market. Then you should schedule your classes according to popularity amongst your student body. Be flexible in this and willing to adjust the curriculum as you observe changing trends in what is more or less popular with your student base.

Private Classes - There are many possible configurations based on just two instructors teaching the private lessons. You can easily schedule fourteen 30-minute private classes per day, split between the two instructors, and still leave yourselves a long break for the both of you prior to group classes at night as well as providing time for a leisurely lunch break. And of course, there are any number of possible configurations depending on your market, your habits and your likes and dislikes. Just remember that you must stay balanced in your energy expenditures to avoid burnout over the long life of your school.

For the average student in your programs, we recommend that you utilize a structure of one-on-one semi-weekly private lessons for tracking progress, maintaining motivation and keeping an ongoing rapport with that same individual. This then will translate to long term retention of that same pupil. And once again, the regular contact with the instructor creates a strong perception of value in the student's mind.

To further bolster your schedule and/or maximize the usage of your facility, you might also consider offering select day classes. This may be viable if there is a substantial amount daylight foot traffic driven by the daytime businesses in your area.

For greater detail on possible scheduling, see the book section on "*Maximizing Your Income Potential*" - Page 45.

Instructorship Programs

There are many reasons you want such a program in your school. And you should take the time to think one through that serves your purposes and then implement it accordingly. Too many school owners tend to "draft" a student who has just completed some part of the curriculum to teach another student that same part. This is a critical mistake. The student who receives their instruction from this far less than qualified individual will tend to feel slighted in this experience. If this happens too often, it will likely cost you a student. The adult professional expects that he or she will receive instruction directly from a certified professional in the pursuit of their martial studies. They will shortly become dissatisfied and leave if they are receiving instruction from "some guy" who enrolled in your school six weeks prior to them. An assistant instructor needs to be well trained before they are allowed to engage in teaching others. This will ensure that you minimize the risk of losing other students in the process.

For one, an instructorship program serves as a potential "perk" for your long term students and again adds to perception of value. It can be *sold* to them as "further education" to refine their understanding of key concepts within your respective discipline.

Have the program form two broad tiers and then as many additional ones as needed. Tier one is initial instruction to prepare a candidate to "assist" a more qualified individual, whether that be yourself in the beginning stages or a more advanced instructor trainee as the program becomes more firmly established over the years. The second minimal tier needs to be ongoing instruction for primary and assistant instructors to both round out the program and to actually cross train your personnel over time.

The second reason to have such a program in place is fiscally related. And this is simply that an established *volunteer based Internship = Paid (to you) Labor.*

Protect Yourself via Contract

You must be very careful how you set this up and this is one place you really MUST consult an attorney who is familiar with the labor laws of your state. They vary from state to state and you must ensure that you are in conformance with any such program you establish. Otherwise, your state's labor board can and will construe such a program as trying to avoid or skirt around such laws. And this can result in massive fines or worse.

Generally, structuring a contractual agreement that designates this person as a volunteer will need to be done. And there must be no form of compensation, up to and including reduction of their regular school class fees, nothing at all. If anything, consider charging these individuals MORE not less for their involvement in your instructorship program.

Look at it this way, large corporations will charge a franchise fee to individuals who wish to learn proprietary corporate business methods to sell their already established and trademarked product line. The franchise fees cover the training, support structure, marketing, financing, the staffing, etc. Franchisees will then work at a facility to be trained initially without being paid. Your trainees will make mistakes that could cost you money and time to correct. And you may also lose valuable students in the process. An instructorship program participant is learning on your "dime" until they are certified and become a productive part of the staff in your school. They will also acquire the "tools" from you to become a professional instructor. They will then be able to apply what they learned from you to earn their own money, start their own school and become your future competitor.

Remember that all instructors you develop ARE the product in your school. And that as owner you must absorb *all* the "product costs" in your school. This will then directly subtract from your profit margin. Once your instructors are certified, they will expect to receive the same compensation as you do, raising your product costs substantially. Consider that trade schools charge substantial tuition for their programs. Trade schools are teaching their students how to "catch their own fish" and feed themselves for life. You are doing no less for all those in your instructorship program and should charge them accordingly. If after being trained, you choose to keep them as part of your staff, you have at least collected some of the costs to be born while training them. If on the other hand, they leave and attempt to start their own structure, you have recouped at least some of the loss that their competition may inflict on your school.

Again, consult an attorney who is familiar with your state's labor laws before embarking on this path. You must make sure that all legalities are taken into account.

The "Non Compete" Agreement

While you are having your attorney draw up your volunteer contract, you should also have him or her structure a "Contractual Non Compete Agreement" so that you may avoid at least one necessary evil of such programs to the best possible level. You must realize that with a well planned Instructorship program, although you are helping to maximize your school's potential, you are also creating and training your potential future competition. So, as a result, you need to protect yourself accordingly.

The law will not allow you to stop them from breaking away from you and teaching in direct competition. But, you CAN have them contractually bound to certain limitations such as how soon they can do so after leaving you, what exactly they are allowed to teach and not teach (copyrighted and/or proprietary methods can often be made off limits legally speaking) or how near they can be to your school (certain limits will apply).

Once your instructorship program is thoroughly established and ongoing in your structure, you should schedule quarterly or semi-annual performance evaluations for the Internship. You should contract all participants in the beginning to such evaluations and then give yourself the room to terminate them from the program for failure to meet their obligations. This way, you can get rid of a troublesome candidate before they become too disruptive to any aspect of your school operations. And for the ones who stay focused, the evaluations will serve as a platform to continually motivate them to improve their abilities.

In the end, if run well, an instructorship program can become a real morale booster for your school and all of the students in it. It can foster a real sense of community and nurturing from older to younger students and generally promote good feelings to the overall population in your organization. The access to gain the necessary knowledge and skills to becoming a capable teacher is a prized thing among many students and the feelings of prestige are very real. This translates to enthusiasm which will be felt as ripples through your entire population. And such enthusiasm translates to success.

Don't, Do NOT, Mix Children With Adults!

This is one of the most frequently expressed complaints among adult martial arts students who have left a school which had mixed children and adults into the same class. They do not wish to be paired with a child in a martial arts class. It is not an ideal learning environment for them and often makes them feel very uncomfortable.

If you DO decide after all to teach children, never EVER mix kids with adults! No matter how much you may be tempted in the start up phases of your school, you must NOT give into this temptation. You will only succeed in driving away your adult population over time as a result.

When asked, many adult students may say that such mixing is fine with them. But they will only be saying what they think you wish to hear. Or, they may be trying not to offend you or your sensibilities since you are supposed to be the professional.

This is a recipe for ruin. Adults and children require different approaches in the classroom. You must not alienate your adult population out of convenience to you or your staff.

If you want to have a kid's program, you must structure and run it completely separate from all your adult programs. Even adjacent "open" rooms where one class made up of children while the class next door is made up of adults is a very bad idea. Children's programs are best kept to their own hours and/or in their own closed off training rooms. Keep them separate, separate, separate, and your entire population will be happier as a result.

MAXIMIZING YOUR INCOME POTENTIAL

There are numerous strategies that can be employed to insure that you receive as much revenue from each member in your school as is humanly possible. You should always be looking to convert any negative "Yin" to a positive "Yang". Be vigilant and proactive in this regard and you will see your bottom line grow steadily.

POSITIVE YANG

Strategy*: Seminars are an avenue for your school's income potential on multiple levels.*

Fiscal: You should aim for conducting one seminar per month. Any usual "down-time" in your school can be used to schedule such activities, meaning when you have any lull in your week. Saturdays or Sundays are often good as students have the availability to attend.

Seminars are a very cost effective way to maximize gross potential profit in your facility. Most schools stand empty during the weekends anyway. You will see the potential additional income which can be generated by perusing three examples of weekend seminar/workshops that follow on the subsequent pages.

Typically, Saturdays and Sundays are not truly popular days for most adult students to attend regularly scheduled classes as they are often reserved for family activities or errands. But most of these same students are capable of attending one weekend day event in each month. And, once accustomed to this, the adult student will actually look forward to them. They are able to learn new things outside of regular class structures and socialize with other schoolmates.

Set a price point for a seminar that includes the cost of a practice weapon, gear, training device or traditional silk uniform for them to keep (depending on what subject matter is presented), and also provide lunch for an all day event. Most professional adult students will very much enjoy a "hands-on" weekend one-day seminar with but a few hours time commitment. It is both fun and intellectually stimulating for them.

Morale: Seminars are a good value for the individuals who are interested in learning new things. It gives students a chance to socialize and generally enjoy the positive experience of the school where they devote their studies.

Most households own at least one home game system, and most of the popular games cost around $50 or more per game at the time of this writing. You should structure most workshops based on that range of cost. This price point will be easy on everyone's budget and allow for a quick commitment in participation.

Seminars can also extend involvement into the classroom. If the seminar is involving a weapon, for example, you should extend the student involvement by also encouraging the attendees of the seminar to come to your "weaponry class" with their new swords so that they may practice the form with their partners. The combined strategy will cause the enthusiasm of the seminar participants to extend to your entire school. This will then further additional student interest in those who wish to enjoy the positive experience during the next seminar. *This is a great morale booster. Word of mouth is the best advertisement for your programs.*

Experiment with different variations of seminars/workshops over time to capture the majority of your students' interest. For example; you might introduce another weapons workshop that includes a higher quality weapon, an aluminum straight sword for instance, and raise the price point accordingly. Adults love souvenirs. Give them one to brag about.

Strategy: An example of a typical Saturday seminar;

Classical Weaponry Seminar. Sword Form and Application, Lunch included. Participants limited to 30 for this sample. Include a practice wooden sword for the attendees to keep.

Charge a fee for the event roughly between 35% to 50% of the average monthly tuition. Fee estimated (based on monthly tuition of $150) at between $55 to $75 per person for the entire event.

Example A: Propose fees at $55 per person, 30 attendees

Projected Gross: $55 x 30 = $1650

Cost: Wooden Sword, estimated cost per single unit is $8 x 30 = $240

Lunch at a nearby Chinese Fast Food Chain is between $5 to $7 per person.

Cost for Lunch: $5 x 30 = $150 or $7 X 30 = $210

Projected Cost for the entire event: $390 to $450

Projected Profit: $1200 to $1260

Projected Profit Margin: 72.7% on every dollar grossed.

Example B: Propose fees at $75 per person, 30 attendees

Projected Gross: $75 x 30 = $2250

Cost: Wooden Sword, estimated cost per single unit is $8 x 30 = $240

Lunch at a nearby Chinese Fast Food Chain is between $5 to $7 per person.

Cost for Lunch: $5 x 30 = $150 or $7 X 30 = $210

Projected Cost for the entire event: $390 to $450

Projected Profit: $1800 to $1860

Projected Profit Margin: 82.6% on every dollar grossed.

*** Caution: Do not be tempted, under any circumstances, to sacrifice potential liability of your school by attempting to cater the event with your own home cooking just to keep the additional lunch profit. This is an extremely bad idea. In an unfortunate circumstance where all of your attendees develop 30 cases of food poisoning you could be out far more money than you think you may save. Your school's liability insurance may not cover the claim. Even if you have a $500 deductible, the out of pocket expense for each case/incident would far exceed the $210 cost of the entire luncheon the restaurant may charge for its services. In addition, the negative experience is irreversible and the attendees will talk. Their "bad publicity" will spread through your entire school effectively killing your weekend seminar revenue stream. Never try to pinch pennies in this instance and let the dollars slip through your fingers. See the section on* ***"Stick To What You Are Good At."*** - Page 26.

<u>Strategy</u>: A second example of a typical Saturday seminar;

Ornate Sword Seminar. Sword Form and Application, Lunch included. Participants limited to 30 for this sample. Include a beautiful metal sword with ornate scabbard (which is also great to display at home) for the attendees to keep. Note: It would be a great marketing strategy to display an actual sample of the sword at your school. Let the sword display sell both itself AND the seminar.

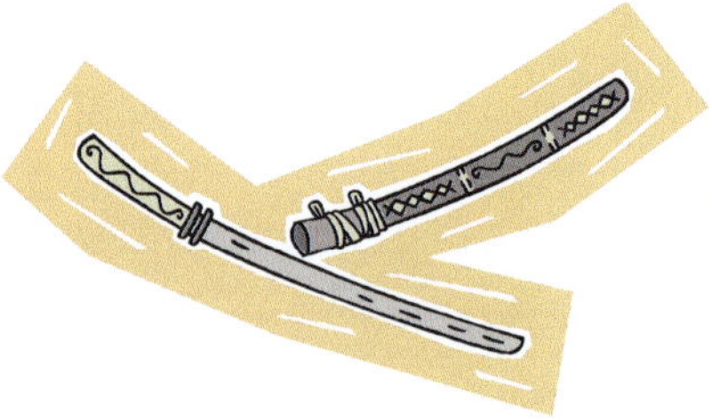

You should charge a fee for this "extravagant" event at close to your average monthly tuition. Fees estimated (based on monthly tuition at $150) at between $125 to $150 per person for the entire event. The suggested retail price for this weapon would have been approximately $120 per piece. Students who are interested in learning the Classical Tai Chi Straight Sword routine will recognize the tremendous value in this seminar. There are also those students who are looking for a more meaningful activity in the weekend as opposed to the usual watching T.V., playing home games, or a seeing a ball game at the local recreation center. These same students would be inclined to attend due to the excellent "word of mouth" marketing that has been forthcoming via the attendees of *previous successfully run seminars*.

Once you have established a schedule of seminars throughout the year as well as established your credibility to host each event with professionalism, the results in generating additional profits to your bottom line will be extremely rewarding. As shown in the samples below:

Example A: Propose fees at $125 per person, 30 attendees

Projected Gross: $125 x 30 = $3750

Cost: Ornate Aluminum Sword, estimated cost per unit is $40 x 30 = $ 1200

Lunch at a nearby Chinese Fast Food Chain is between $5 to $7 per person.

Cost for Lunch: $5 x 30 = $150 or $7 X 30 = $210

Projected Cost for the entire event: $1350 to $1410

Projected Profit: $2340 to $2400

Projected Profit Margin: 62.4% on every dollar grossed.

Example B: Propose fees at $150 per person, 30 attendees

Projected Gross: $150 x 30 = $4500

Cost: Ornate Aluminum Sword. estimated cost per unit is $40 x 30 = $ 1200

Lunch at a nearby Chinese Fast Food Chain is between $5 to $7 per person.

Cost for Lunch: $5 x 30 = $150 or $7 X 30 = $210

Projected Cost for the entire event: $1350 to $1410

Projected Profit: $3090 to $3150

Projected Profit Margin: 70% on every dollar grossed.

Coming up is yet another set of examples regarding how you can and should use seminars to take advantage of seasonal sales that stem from your martial arts wholesale suppliers. See the following for two examples of seminars on the *Softer Side of Studies:*

Tea Ceremony. Consider offering a Tea Ceremony to portray and educate on the softer side of traditional Chinese customs to enhance the experience for the seminar attendees.

ALL martial arts students embrace things that are related to the culture of the Chinese Martial Arts. You should capitalize on that. The offering of a Tea Ceremony experience to your students will provide an up-sell opportunity to your Chi Kung and Meditation workshops. Instead of the typical lunch fare offered in your other seminars. You can include a souvenir ornate tea cup (decorated with classical Chinese painting of Tiger, Dragon, Phoenix, Crane, Pine Tree, Plum Flowers, etc.) which can be purchased at Asian gift shops in most China Towns or online. Charge a fee to cover the cost of the tea, the cup and props. The tea ceremony will also add length and depth to your seminar which translates to perception of value. If this is not an area of your expertise, rectify it by investing some time in reading up on the Tea Ceremony on the internet or pick up a book on the subject. Showcase your well-roundedness in Asian Culture beyond your family style of martial arts. Be creative, enrich the five senses of all attendees. Create the ambience of this type of seminar with music and some tasteful decor in addition to the flavor of Green Tea and the aroma from burning sandalwood incense, etc.

Example A: Propose fees at $100 per person, 30 attendees

Projected Gross: $100 x 30 = $3000

Cost: Satin uniform top and bottom, estimated cost per set is $30 x 30 = $ 900

Tea Cups, Green Tea and props $5 per person.

Cost for Tea Ceremony: $5 x 30 = $150

Projected Cost for the entire event: $1050

Projected Profit: $1950 to

Projected Profit Margin: 65% on every dollar grossed.

Chi Kung and Meditation seminar. Silk or satin traditional uniforms can be a great compliment to your Chi Kung and Meditation seminar. You should provide a softer side of study to capture those students who are more timid and are interested in solo routines.

The estimated wholesale cost, at the time of this writing, for each set of satin/silk traditional uniform ranges between $30 (satin) to $50 (silk), suggested retail price is between $75 (satin) to

$135 (silk). You can offer this seminar ranging between $100 to $155 dependent on including either a satin or silk traditional uniform. The profit margin of this seminar will be between 64.5% and 65%. Students do not normally have a reason to purchase such merchandise from display in your retail shop alone. Yet you still carry the inventory costs, not being able to turn a profit on such items. Regularly scheduled seminars can give you an opportunity to link such merchandise with purpose, yielding greater potential profits AND building enthusiasm within your school. This will in turn promote the longevity of your martial arts business.

Example B: Propose fees at $155 per person, 30 attendees

Projected Gross: $155 x 30 = $4650

Cost: Traditional Silk uniform top and bottom, estimated cost per set is $50 x 30 = $ 1500

Tea Cups, Green Tea and props $5 per person.

Cost for Tea Ceremony: $5 x 30 = $150

Projected Cost for the entire event: $1650

Projected Profit: $3000

Projected Profit Margin: 64.5% on every dollar grossed.

Seminars are also an effective "Test Market" for your new curriculums. They provide a great opportunity for testing possible new technical classes like Knife Defense, Chin Na, Combative Arts; or softer studies such as Meditation, Chi Kung, etc.

You might also consider bringing in *outside talent* occasionally. Someone who is not of your particular stylistic persuasion but has something of value to offer your student base within a seminar environment. "*Diversity Sells*".

Strategy: *Produce Videos; even if only for your own in house students and sales.*

Match your ranking structure precisely with each rank having its own video. Professional adult students are accustomed to utilizing training manuals/textbooks in their profession. Video will be a natural aide to their martial arts endeavors. Consider giving them the first DVD free when they become a member as part of their "startup package". They will learn to appreciate the effectiveness of such a tool in the beginning of their involvement with studies. And then, after that, they will have a greater tendency to purchase additional DVDs as they progress through the ranks.

This is a “Win Win” situation for everyone concerned. You win with increased sales. They win with increased tools to aid them in their learning curve. Establish an effective “up-sell” strategy to firmly put the notion in your student’s minds that they both *want* and *need* the videos. Get this point across early and keep re-enforcing this perspective in them.

Wisdom*: Tread carefully with Testing Fees.*

As you structure your curriculum, embedding it with small victories/goals for your students to achieve and thereby stay on track with regard to their progress, ***do not charge fees for minor tests!*** Such minor achievement tests can occur during their semi-weekly privates, then denoted by giving them a small patch that can be sewn on to their sashes acknowledging that progress to their peers, for example. This way, they get the bragging rights but also privacy in the exam. Incorporating such minor tests into the private lesson effectively allows you to charge for the time taken in the exam, but it doesn't feel like it to them which translates to perceived value.

Charge only for the more important tests that are considered monumental to your structure such as those which include a new colored sash that signify a complete rank promotion. This makes these types of exams feel really special and you will re-enforce that within your school’s collective mind. You should also consider showcasing more gifted individuals having them test in front of the entire school for motivational purposes. But, by the same token, a less than gifted individual testing in front of the class or school can work against you in terms of morale.

Strategy*: Annual Membership Fees are a must but proceed with caution on how you sell it.*

Annual membership fees will help cover expenses in your school’s added amenities such as equipment, books for the library, newsletters, signs, etc. They also help to offset the 3% annual inflation costs of operating your school.

An annual membership fee, while a must for your school, can feel like “money, money, money” to your students. And you do not want to send that message ever. So, disguise it to make it more palatable to all concerned. For example: Most students do not purchase additional uniforms unless they come to classes more often than they are able to do their laundry, or, if the uniform gets ripped, worn or faded.

So when they renew their membership, consider including a new uniform, perhaps of a different color scheme, trim on the collar, sleeves, pants or even a patch that denotes their tenure to the school. This will make other students tend to covet the same uniform and some will want to stay simply to get the same uniform. This will also impact the appearance of your student body in a positive way. Less students will come to classes with beaten up uniforms.

Your students will be more positively inclined to pay the annual membership fee if they know they are getting something in return. So give it to them.

Strategy*: Each student should have at least one private lesson per month, preferably two.*

Always be sure to assign an instructor with each student from the beginning of their study. It is a must to schedule the first private lesson as "Orientation" where the student will be exposed to the fundamentals prior to attending their first group class. This will help ease the beginner's "First Day Jitters" and doing so can be crucial to whether or not the beginning student returns to classes after day one. Having regular private lessons will allow you or the instructor to be able to establish a rapport with the new student during his/her regularly scheduled private lessons.

Always be sure to assign an instructor with each student from the beginning of their study.

Ideally, you should consider offering one private lesson every two weeks. This will allow you to keep each student on track with their studies. If you absolutely do not have the time in your normal schedule to provide two private lessons on a monthly basis to each student, then a once a month frequency is better than nothing.

You must make no mistake here. Private lessons are an absolute must for the development of a cohesive program. This synergy of private training combined with group training will not only yield progress in the student but the creation of such a program will translate into tremendous student retention for your school. It is an absolute must to construct your programs accordingly and to allow time in your day to accommodate such programs.

Following is another example of a private lesson schedule based on two full time instructors:

In the below example both instructors simultaneously conduct the lessons and the two of you will be able to accommodate 14 private lessons per day, 70 students per week, 300 lessons per month. This will allow for each of 150 students having two private lessons per month with their

Proposed Business Hours:	Monday through Friday
	2:00 pm to 10:00 pm
Structure Private Lessons at 1/2 hour each:	
Example of Private Lesson Schedule:	
2:00 pm to 2:30 pm	
2:30 pm to 3:00 pm	
3:00 pm to 3:30 pm	
3:30 pm to 4:00 pm	
4:00 pm to 5:00 pm	*Break/Conference* (see below)
5:00 pm to 5:30 pm	
5:30 pm to 6:00 pm	
6:00 pm to 6:30 pm	
Group Class Schedule:	
6:30 pm to 7:30 pm	
7:30 pm to 8:30 pm	
8:30 pm to 9:30 pm	
10:00 pm	*Close*

instructor. This is an ideal program for all students on a regular basis. The every-other-week interaction with each student will yield great results in each student's progress and help to maintain positive student relations. Most of all the student will feel like they are being treated as an individual. You should encourage all students to also attend group classes weekly. Then you can utilize the private lessons to further enhance their understanding, rank testing, coaching and guidance to keep them on track.

Use the scheduled *"Break/Conference"* daily for what it is intended. It is vital to your school's operation to maintain constant communication between instructors. This allows the time to strategize, update and keep each instructor on track in relation to what goes on in your school and student body in general. Your school is a living entity, you should nurture it as such.

<u>Wisdom</u>*: Keep a "Performance Tracking Log" in each individual student's file.*

When was the last time that some professional service provider mistook you for another one of their clients? Or, they *misplaced* your file and did not know the reason why you were there for your appointment. It happens to all of us from time to time. We tend to just shrug it off and

think "Well, they do deal with a lot of customers and clients on a daily basis. And it is bound to happen sometimes." But it didn't make you feel good did it? And this is a very bad feeling for your students to have as well. And once the negativity is formed it becomes difficult, if not impossible, to erase. This is not the ideal result you hope for within your school. See the section on *"Maximizing Each Students' Gross Potential"* Page - 61, to realize how much revenue is lost whenever one of them quits.

The "performance tracking log" in each of your students' files should include categories such as: current areas of focus and challenge, long term and short term goals, future review topics and review dates, and most importantly their individual purpose of enrolling in your school in the first place. Not only will this ensure that you remember their names, but also any other vital information about that student which helps you treat your students as individuals and keep that ever so important rapport growing.

By developing such an individual performance tracking system, you will enable your school to deliver the highest quality of service within your institution of learning. This is an absolute MUST in providing the optimal learning environment for your student body. This particular practice in student management will reflect your professionalism in the eyes of your students and leave a lasting impression as to your level of caring with regards to their progress.

It is a reality that once your student base has grown larger it will become harder and harder for you to keep track of performance status on each member. Consider this to be a *good problem.* Manage it meticulously to ensure the *good problem* does not turn bad on you. Never should you allow this practice to slip or become something you do occasionally. You must find ways to maintain this practice no matter what. This is vital for the long term health of your school.

Here is an example of a simple "Student Performance Tracking System":

Name:	Brian Daniels
Reason for Enrollment:	Original reason(s) they came to your school
Today's Date:	March 15
Current Area of Focus:	Better side kicks
Current Area of Challenges:	Balance
Immediate Short Term Goals:	Learn the techniques to advance to the next Belt
Long Term Goals:	Improved overall health, Stress Management
Future Review Topics:	Refine required forms to the next Belt
Future Review Date:	March 29
Expectation:	Graduate to the next rank by the end of April

The above example allows you to be updated on the individual student's status with just a glance at the beginning of the private lesson. The brief notes made in each category take only minutes to complete at the end of the private lesson. The concept is to make the process simple and concise and then for the instructors to adhere to it as a constant. Start by implementing a form such as the example shown above (or modify to suit your program) on day one during "student orientation". You and your partner instructors *must* be properly trained on <u>how</u> and <u>when</u> to enter the information in each category. Otherwise, you will commit the cardinal sin of a cliché *"Garbage in. Garbage out."*

<u>How</u> to fill out the form? Simply go down line by line and write a brief and succinct comment in each category entry. It must make sense at a glance.

<u>When</u> to fill out the form? Preferably immediately after the end of each and every private lesson with their assigned student.

Why immediately? If the information entered is not accurate, you might as well have not done it at all. By the end of your day, the likelihood is that the information retained in your memory will not be accurate. Your recollection of the specifics of those particular students you have taught during that day will have likely become more of a "blur" and you will then "fudge".

If you are going to take the time in doing a task that will improve your student retention, make it count! Do it right! Your students will appreciate the effort put forth by your instructor staff to track and care about their respective progress. You will demonstrate *respect* and *consideration* towards the growth and progress of each and every member in your school. And this will in turn translate to loyalty, through perception of value, in their minds.

IN-HOUSE SALES

The working credo of a successfully run business should be to keep as many services in-house as possible. Corporate structures have known how effective this for decades and you should too.

"In-House" services allow greater quality and cost control which translates to happier customers. And cutting out all "middle men" wherever possible, "in house" translates to a higher profit margin for your school. A martial arts school provides instruction yes, but should also provide *all* amenities that are relevant to your students' studies. You should aim to supply the *needs* of a martial arts student. Keep everything under one roof and offer all conveniences on your premises so that your students will not need to go outside of your school to satisfy their product demands.

Therefore the essentials of a martial arts business must include not only the monthly tuition of group classes and private lessons, but also must include retail sales, seminars, testing, tournaments, etc. all in addition. It is imperative that a martial arts instructor/operator provide easy access for all students to participate in *all* aspect of your services.

Strategy*: Auto Debit is your key to success.***

The old days of collecting payments from your students by personal check, which they either mail in or bring with them when they come to class, needs a serious overhaul. Oftentimes (if you still use this method) many payments will be late which affects your cash flow. In addition, you will also have to spend time preparing billings for any past due accounts. If you are still allowing such methods of payment, you should strongly consider upgrading the method you use for collecting all accounts receivable.

Auto Debit or Auto Pay is commonly used by most service industries that collect recurring charges from their clients either monthly or quarterly. Using auto debit eliminates the labor hours spent in processing billing and collecting accounts receivable. You will immediately experience an ease in cash flow once *Auto Debit or Auto Pay* is established in your school.

The best time to make such an arrangement is during the "sign up orientation". All your new enrollees should select, based on their personal preference, a fixed date of either 1st day or 15th day of each calendar month to have their monthly fees debited. Your school will then assume the responsibility to debit their account monthly, providing the convenience to your students of not having to keep track of writing checks every month. This will save you the headache of waiting for students to mail their check in on time. The cost of such a policy is merely a minor credit card transaction fee (which can be passed on to your students within their tuition fee). The benefits to you are better time management and greater control of your cash flow.

Strategy: ***Credit Card acceptance is the answer to your cash flow.***

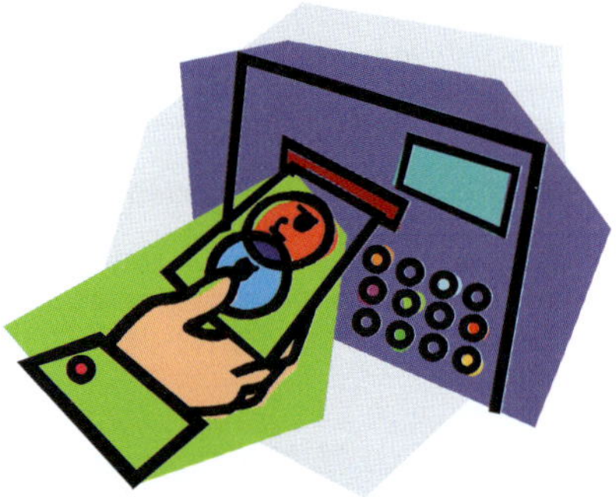

Virtually everyone has credit/debit cards that they use regularly these days. All modern commerce has been trending towards cashless transactions for some time now and you should keep up with the changing trend.

Talk to your bank about setting up Merchant Card Services in your facility as soon as you are ready to open for business. The convenience of this arrangement will save you hours per week in time normally spent running bank errands. You can use this time savings to be more productive. Spend that same time in your facility signing up new students or creating new marketing plans and strategies and promotions, or structuring new study programs. You could even use this time savings to take on additional private lessons and generate additional revenue.

Wisdom: ***Here are some facts regarding the acceptance of major credit cards:***

Cash Flow- You should set up your Merchant Card Services through a national bank who can process your sales directly, and credit the sales to your checking account within the very next business day. This means the funds will be available for you to spend immediately. There are other types of processing centers that route the transactions through third party brokers and/or agencies. This may translate to a delay of several business days in crediting your accounts. This delay can literally cause an accounting nightmare for you to deal with on a regular basis. And this same delay can also put a "crimp" in your cash flow because you are waiting for your money to hit your account. These same agencies may entice you with fees that are less than the average, but you should weigh everything carefully before deciding.

Transaction Fees– The average typical transaction fees imposed by major national banks are around 4% to 6% (at the time of this writing) per every dollar processed through your credit card terminal. The rate selected will depend on your average transactions, which are posted through your credit card terminal. *The higher the average transaction the lower the rate or fee will be*. The fees *can be* negotiable. All you will need to do is thoroughly research what their competitors are charging. Then ask your bank to match the competitors rate.

Merchant Services will conduct an evaluation of your average transaction routinely every six months to a year. You can request to have the evaluation done every six months, if desired, based on the business trend in your school. You would want to make such a request in hopes of reducing your rate due to a higher average transaction, if this is the trend you see.

Auto Pay Program– Once you have Merchant Services in place, you should immediately set up an auto pay program for all new students' monthly membership fees, to be enacted upon them enrolling in your school. Make sure to have every new student sign a proper release or authorization according to the card issuer's agreement. Your Merchant Services should include such literature in your welcome package when you set up your credit card terminal. If not, you should contact the Customer Support of your Merchant Services agency.

Stimulate Sales- Students who are in an *Auto Pay Program* tend to be happier because they essentially have an account on file with the school. This lets them purchase retail items at any time, sign up for weekend seminars or pay for their tests, etc. All without bringing their wallets to the school.

Meanwhile, your school gets paid in full at the point of sale, which will improve your cash flow immediately. And you have the powerful sales tool of simply saying "would you like to put that on your account?"

Your students will be less likely to hesitate in purchasing large ticket items because they can simply use their credit card on file. They will be more easily able to justify such purchases because they can make installments to the credit card over the span of a few months. Meanwhile, your school gets paid in full at the point of sale, which will improve your cash flow immediately. And you have the powerful sales tool of simply saying "would you like to put that on your account?" The ease of credit cards ensures the ease of sales, whereas you may have otherwise lost the sale if you chose not to accept credit cards in your school.

Tele-Check/Electronic Debit– This service will allow you to convert any written checks to electronic checks by posting them through your credit card terminal. You will be able to verify that the checking account has the funds to cover the amount and then instantly debit the account the check was written from. You can prevent bounced checks and any "Stop Payment" orders on checks written to your school by setting up this service when you set up your initial credit card terminal.

There will still be some students who prefer to pay by check. Consider charging an additional $5 to $8 per month for checks written to encourage them to sign up under *Auto Pay Program*. You have a cost for each check transaction. Pass that cost on to your students. Students who prefer to pay by check also tend to be the same individuals who are late with their monthly fees. Impose a late fee penalty on the offenders. These same individuals could also be those who are asking to make installments on large ticket items such as sparring gear, weekend seminars and monthly dues. Our experience is that the risk and headache outweigh the benefits of making that sale. When you allow a student to pay in installments, the first installment will likely not cover the cost of the item(s) in question, so you are then already at a loss on that item or items. Furthermore, allowing a student to pay in installments will cause you the hassle of preparing extra paper work to track the payment history on said item(s) costing you additional time. And in the end, you may not receive the payment in full.
Remember that, after all, plastic does not "Bounce".

<u>NOTE</u> - You should get very acquainted with all the laws governing Credit Card Sales in your state by thoroughly reading the material provided by your Merchant Services. Becoming familiar with the features and benefits will enable you to be creative about utilizing your Merchant Card Services to maximize the gross potential of your school.

Never allow anyone other than you and/or your business partner to accept any payments from students.

<u>Wisdom</u>*: Only you (and your partner) should maintain complete control of all "Receivables" once you set up the automated payment program in your office.*

Never allow anyone other than you and/or your business partner to accept any payments from students. If you do use someone else, they must be well trained to follow the proper protocol. Designate either one or both of you only (if you do not have an office manager) to sign up new students, post sales and handle all financial matters. This is the best method to ensure that things do not slip through the cracks due to a lack of communication, or worse, mistakes made in charging to the wrong credit card accounts. These are the types of human errors that could cost you students over time. Your own pride in ownership will prevent mishaps in this area from occurring. But don't expect an hourly employee to be as attentive.

<u>Wisdom</u>*: Maximize Each Student's Potential*

"It is easier to keep an existing client/student than to gain a new one." The wisdom in this saying is indisputable. So first, you must understand how a martial arts school generates its revenue. This understanding will then help you realize the importance of "student retention". We have provided samples in this chapter that demonstrate the projected gross potential for each student.

<u>Source of revenue: Tuition</u>

Students are there to learn and be a productive member of a well run martial arts school. You must understand that your student body is a "captive audience" within the walls of your school. On the average, each student will stay in your programs for about two years, more or less, and so you can project two years of tuition revenues on each student on average. Longer would, of course, be better.

<u>Source of revenue: Annual Membership Fees</u>

You should charge an Annual Membership Fee for the first year in which a student enrolls and for every year thereafter in which they are a member in your school. Such a fee will help to cover the amenities you offer, such as equipment, library, etc. which benefit your student body and translate to better cash flow for you and your school.

<u>Source of revenue: Retail Shop</u>

Keep everything in-house regarding your students' equipment purchases. All training uniforms, patches, supplies, training gear, practice weapons and/or videos/DVDs. A well stocked retail shop is a convenient, time-saving amenity for your students. Your average new student will need to acquire training uniforms (top and bottom), a school t-shirt with custom logo, training shoes, a belt or belts (depending on your structure), patches, water bottle and a sport bag to hold everything before they attend their first class. Many of those items are only available through martial arts supply stores in your city or online.

Busy, professional adult students would rather not spend the time running all over town for these specialty items if they can avoid this time drain. The common rule about business is: "*Supply and Demand"*. Your new students *Demand* to buy, you should seize such opportunity and *Supply* these items in your retail shop so that they can make an immediate purchase and charge such items to their credit cards. Your own in-house retail shop, if run well, will yield a substantial additional profit to your bottom line on a consistent basis throughout the year.

Source of revenue: Seminars

You should estimate 1 in every 3 students will attend at least 50% of your seminars conducted throughout the year. Use this average to calculate the "gross potential" of each student in your school with regard to seminar attendance and the extra capital generated there.

Source of revenue: Testing Fees

You should plan and allow for at least one test per student every three months. Too long of a gap between tests and/or evaluations may cause the student to lose focus. On the contrary side, testing too frequently, for example every two weeks can have a negative impact on a student's awareness. It can make the adult feel like they are being *micro-managed* which is a tactic that is suitable for teaching children only and will be recognized as such.

Here are some sample breakdowns of "per student" revenue potential:

Example A: Based on monthly tuition of $150.

Tuition	$150 x 24 months	= $3600
Annual Membership Fees	$ 99 x 2 years	= $ 198
Seminars	$300 x 2 years	= $ 600
Retail Purchases	$250 x 2 years	= $ 500
Testing Fees	$100 x 2 years	= $ 200
Projected Total over two years per student		= ***$5098***
Projected Average Monthly Fees collected per student	=$ 212	

Example B: Based on monthly tuition of $100.

Tuition	$100 x 24 months	= $2400
Annual Membership Fees	$ 69 x 2 years	= $ 138
Seminars	$200 x 2 years	= $ 400
Retail Purchases	$175 x 2 years	= $ 350
Testing Fees	$ 70 x 2 years	= $ 140
Projected Total over two years per student		= ***$3428***
Projected Average Monthly Fees collected per student	=$ 142	

Having seen the above examples, you should now more easily be able to envision your school's potential of being a successful martial arts business. And one that can generate a great deal of cash flow on an annual basis. The potential is truly limited only by your work ethic and your ability to capture market share.

While being mindful of the financial potential, also remember always to invest regularly in your school's services:

1) Operate and outfit your school with modern technology/equipment.

2) Offer outstanding services to your students.

3) Spend your advertising dollars wisely.

4) Recruit quality individuals to be in your structure.

5) Keep each and everyone of your students happy for as long as possible.

Raise Your Rates <u>Annually</u>. *This is a MUST once your school is established.*

Raising rates regularly and incrementally will ensure continued success of your school. You do not want to wait until years have gone by without rate adjustments and end up in a state where you are now having trouble paying your monthly expenses. The result can be catastrophic!

Let's suppose that your school has gone ten years, or since your grand opening, without any rate increase. Meanwhile, you have been gradually depleting your cash reserves over the years due to the inflationary impact of your fixed overhead. Let's suppose that inflation in your area is 3% per year, multiply by 10 (years). You would need to raise the rates by 30% or more immediately just to stay in balance fiscally. Doing so without making any changes to your program to balance or justify the rate increase in your students' minds, and you may experience a mass exodus in your student body as a result. You may lose students that you have invested years in teaching. All the efforts spent to keep them in your programs, gone with a single mistake.

As the old saying goes, "*It is easier to keep an existing client than to gain a new one.*". Such an oversight may begin the demise of your school, that which you have poured your blood, sweat and tears in building for several years. Don't panic if your school has been guilty of not keeping up with the growing economy. You can benefit from a program and pricing structure overhaul. Read on for strategies that you may implement immediately to bring your school to prosperity.

Strategy*: Begin with matching the Cost of Living at a minimum.*

The bench mark for incremental rate increases should be based on the annual increase in Cost of Living of your respective state/county/city, *not* whether your competitor schools raise or lower their rates. Each year, all companies across the world routinely raise their fees and pricing structures due to the inflationary factors that occur throughout the globe. Regularly occurring annual inflation results in the increase of your standard overhead costs such as rent, utilities, accounting, advertising, insurance, inventory, material, repair and maintenance and "business to business" service fees. This in turn narrows your personal profit margin if the annual gross revenues remain the same. You will see the effect in that your bottom line reduces over time and your cash flow suffers.

Don't wait! Do this routinely or it will start a slow death commencing in your school. Try to make the increase nominal and incremental. If necessary break it into "installments" throughout the year. For example; two nominal increases in an amount that does not affect anyone's budget drastically, as opposed to one larger increase may be more palatable. Your students will be more receptive to such minor and necessary adjustments to ensure the survival of their beloved school in such a circumstance.

Wisdom*: Don't be concerned with potential defection of a few older students.*

This is just the nature of any business when it comes to rate increase. It is simply a reality that you can not make everyone happy all of the time. The resistance comes from the simple fact that such an increase will directly affect your clients'/students' pocket book. Expect that there will always be someone who reminisces about *The Good Old Days*... when the dollar used to stretch further than the present time. It is best you just let them go. Such "*Dinosaur thinking*" will inevitably affect the mentalities of other individuals in your school. *And if they stay, they will possibly poison the rest with their negativity. Let them go.*

Why should you and your school struggle financially just to make such individuals happy? Those people will be and should be replaced by students who are both more productive and supportive of your school's endeavor. You should combat any such objections to a rate increase by providing better service to your students year over year so that they will not even want to consider abandoning their studies in your school. No one wants to be a "quitter", so don't give them the reason to be one. If you continually provide better service, they will not want to quit simply because of a nominal rate increase. They have already invested time and effort in achieving their respective ranks within your structured curriculum and they will want to stay.

Strategy: ***Offer additional services; such as private lessons and/or extra curricular activities along with new pricing structure when implemented.***

If your school has been offering only group classes since the beginning, then you may consider adding programs that include private lessons at various frequencies such as once weekly, semi weekly or monthly.

Expect that most of your student body who have been in a group class program will not upgrade initially, simply because they will want to keep their tuitions low. However, those students who are more devoted to their progress will embrace the idea of having ongoing private lessons with an instructor on a regular basis. (**See the example of the tier program list, Page - 79.)

You can offer all new enrollees newly revised study programs that include group classes as well as regular private lessons. At the same time, you should eliminate any and all "group classes only" programs systematically. New enrollees will select a new program which is now inclusive of both private and group classes. They will be sold on the outstanding curriculum and this "hands on" approach to their studies. Your school will now begin with a "clean slate" with every new student as they were not present to know the "history" of any such lower cost "group classes only" programs in the past.

You must proceed cautiously with regards to approaching your existing student body with any change that results in the increase of their monthly tuitions. You should always keep in mind the importance of giving your student body "*Perception of Value"* in exchange for any kind of rate increase.

Your students are there to learn. So give them what they want by offering unlimited attendance and adding some new classes to balance the cost increase.

You may also consider holding annual school tournaments to add value to your programs. The cost of such events can be kept down to a bare minimum simply because the event is being held in your school during regular business hours. Purchase token trophies and/or medals from your suppliers during their own seasonal sales during the year for use at such events later. You can also present personalized certificates generated by your computer and printer. Such events are great morale boosters. Everyone likes to be acknowledged with regard to their hard work and praised for their outstanding performance. Tournament events work well for building mass enthusiasm and camaraderie within your school. You should also consider recording the event and then adding these annual events to your DVD archive for both display and viewing in your student library. The result of your minimal investment will yield a ten fold return.

Wisdom*: Be consistent with your rate adjustments across the board.*

You should also consider adjusting existing fees for testing, seminars, private lessons, retail products, etc. accordingly and in keeping with the cost of living increase. Once a routine annual rate adjustment is implemented, your students will be educated from the beginning about the reality of annual rate adjustments. They will accept it as the *"norm"* in your school.

Strategy*: Capitalize on the tendency of a surge in sales and enrollment before each "Annual Rate Adjustment".*

Allow yourself to tap into the mentality in consumers who attempt to seize the *"right time"* to make their purchases. This can be a significant side benefit of the "Annual Rate Adjustment" event.

The reasoning is sound and simple: People naturally seek *"bargains"*. Many consumers will wait for a *"sales event"* to make a purchase. Most people will attempt to maintain a perspective of being *"sensible"* as opposed to *"impulsive"*. Therefore, a regularly scheduled "rate increase" will prompt some of them to make *smart* decisions in order to *take advantage* of the lower rate or price tag prior to the "rate hike". Your prospective members and student referrals will be more willing to sign up for a long term programs at this time to *"lock in"* their rate before the "rate increase". Some of your existing students will tend to make purchases on items that they *have been eyeing* just before the price tags of your retail shop go up so they may save a few dollars.

Let's face it, saving money and getting a great deal presents a positive feeing in people. You will have created a "win win" situation in implementing the "Annual Rate Adjustment" in your school. The consumers *"win"* in getting the *"great deals"*. The revenue stream in your school grows as a result. Your martial arts business *"wins"* either way.

Maximizing Your Income Potential

NEGATIVE YIN

It is likely that you have tried many promotional offers in your school by now. Many of which have not yielded the results you had hoped for. Some school owners will often try to undercut what they see their competitor schools are doing by lowering the price on promotional offers. Before you do such a thing, make sure you gather all necessary information so that you can make an informed decision. You need to have all of the facts and details of your competitors' promotions such as free gifts, classes and/or private lessons offered by your competitor schools. Calculate all costs very carefully involved in both yours and your competitors' promotions before entering a "Price War" against your competitor schools. If you fail to heed this advice, the result will no doubt hasten the depletion of your cash reserve and crimp your cash flow.

It is, of course, necessary to experiment with different promotions to stimulate interest in new prospects. The goals are to generate income and increase your student base. It is crucial to the success of your school to not make mistakes in this area.

So how do you avoid pitfalls when trying to market the right promotional packages? What is the proper price point? How do you go about constructing the perfect promotional offers and give your prospects the incentive they need to join your school? Here are some steps to take in developing your arsenal of effective promotional packaging:

First: ***What is the profit margin for each respective promotional offer?*** You must first calculate the cost of the merchandise and/or services in the promotional offer to determine your profit margin on that respective offer. Then you must structure the promotion at a price point that will entice prospective members to *snatch this awesome deal* and join your school.

Second: ***Identify any promotional offers that do not work by compiling daily data.*** It must be expected that certain promotional offers may not have the impact you had intended. To move toward a workable solution, establish a data base of all promotional offers. This data base should track the number of the appointments and new enrollees each promotion generates since the day of its institution.

Such a data base will allow you to gradually acquire a historical record of your sales trends and marketing strategies at a glance. If possible, create a category to track any voiced objections you hear from prospective members with regards to why they are *not ready* to enroll today. Voiced objections could be things like the price point; undesirable length of program (too long or too short); the number of private lessons offered; the merchandise offered; or even the timing (summer vacation, holidays, academic commitment, etc.). Seasonal factors can greatly impact the success or failure of any promotional offer. This is something you must not ignore.

Third: ***Make adjustments to transform the Yin to Yang, Negative to Positive.*** Once you have successfully identified the reasons of any non-effective promotional offers you will be in a better position to structure new and improved promotional offers that can more effectively market your school. Utilize the data which you have gathered from all objections received from prospective members who did not sign up in your school. Adjust the price point according to the disposable income levels of the market share you are after. Roll out seasonal promotions, using appropriate buzz words or catch phrases in the title of the promotion, that will resonate in the general populace in your area to entice new enrollees.

Fourth: ***Establish rotational promotions to match the seasons and trends.*** It is quite important to run various promotions in your marketing strategies. For instance, your New Year's Resolution promotional offer will not have impact in the summer. For Summer, you might instead construct a two months trial study course to entice college students during their summer break. And of course a holiday seasonal promotion will not be as effective in times other than the holidays. Match your promotional offers to seasons, holidays and other trends in your area. You might create promotional packages in accordance to the current health trend for example. Or, you might market personal empowerment programs to professional groups or clubs. Additionally, expand your marketing strategies wherever possible to successfully reach the clientele you desire by being both creative and innovative. (**See the sample promotional offers - Addendums Two - the end of the book - Page 6)

Fifth: ***Execute properly to maximize income potential and student enrollment.*** Train your sales person on each promotional offer so that he or she may properly represent your school. You want all prospective members to have a good first impression of your school by experiencing confidence in your knowledgeable personnel from day one. There is nothing worse for your school than uninformed, poorly trained personnel. This will be seen as a lack of professionalism in your representatives. The prospective member will leave your school with a conclusion that yours is not the school for them. You cannot at all afford to have this happen too often. Though it is un-realistic to have a 100% sign-up rate (signing up ten out of every ten prospective members month after month) you should aim to achieve at least an 80% sign-up rate, or signing up eight out of every ten prospective members consistently. This will ensure a continuous growth in your student base which will result in greater income potential.

THE HIDDEN COST IN "DISCOUNTS"...

There are costs associated with every "discount" offered. You will be out of pocket the wholesale price in each "free uniform" given away. For every promotion that offers "Free Trial Classes" you are out of pocket the standard overhead per student (day to day costs of running your school). "Free Private Lessons" will cost additional labor dollars on top of the standard overhead costs per student. These costs are tangible and they should always be controlled. Use discretion in such instances.

However, there are other "costs" in certain types of discounts that are less tangible and hence more treacherous to navigate. Here are some examples which should cause a *"pause for thought"* before such things are considered for implementation:

"Family Discounts" is a common promotional package you will find in most fitness clubs, yoga studios, dance studios and martial arts schools. The intention behind such a promotion is to bring in additional family members-by-association to your school and boost your bottom line as a result. Great idea. Right? Not necessarily. This is why...

1. When you offer discounts to *"family members"*, you are unable to realize 100% of the "gross potential" per student as shown in the examples from ***"Maximize Each Student's Potential"*** section - Page 61.

2. Too many *"family members"* in your school will limit your capacity to accommodate other average students who will be there for their own unique purposes. These students would be paying you 100% of what you charge for your services. But his or her space is instead taken up by a family member who is paying you a discounted fee.

3. <u>*The most important criteria in growing your membership base is retention.*</u> And the average solo student has a much greater potential of staying in your programs longer than a *"couple"* or a *"group"* as discussed in the following pages:

Definitions of ***"Family Discounts"*** include the following:

"The Spousal Discount" It may seem like a good idea on the surface at first. But, not so. The *"couple"* will take up the same space in your classroom as two regular students. Your time and attention required to teach them will be the same as any other two individuals in your school. Yet, you get paid less to teach the *"couple"*. In most cases, only one of the spouses is the "initiator", meaning the one who is truly interested in the studies. The other spouse is there to be "supportive" and so that they may do something as a *"couple"*. Sooner or later, inevitably, the less motivated partner decides that he or she no longer wishes to be in your program. This often causes both of them to quit simply out of a feeling of obligation to their partner. There usually is nothing you can do to salvage them. You end up losing two students.

"The Sibling Discount" The dynamic of the *"siblings"* association is very similar to the *"spousal"* association. You may find that only one of the siblings who joins your program is truly motivated. The other or others are motivated only by the enthusiasm of the "initiator". Again, this generally will not be productive. And eventually you will end up losing both or all of them when one decides they have had enough.

In addition, at times you may encounter certain individuals who would try to take advantage of the discount by stretching the definition of ***"Family Discounts"*** to the following:

"The Cousin Discount" This is one of the most abused discounts, often falsely claimed by friends of your students' looking to get a break on their monthly tuitions if you allow it in your school. It is simply bad form for you to ask for proof of this "association" or "filial/family" tie in this instance. Yet, you will feel obligated to honor the discount, regardless. You again will get paid less to teach these people. Too many of these "discount" students in your school will cut into your bottom line.

Wisdom: ***Offering "Family Discounts" is a mentality that is counter-intuitive to both your students and your business.*** First of all, your students are the reason you open your school. You must realize that your students endeavor to better themselves through the involvement of martial arts studies. It is an involved and personal pursuit to reach their individualized goals. Your students expect that the school they involve themselves with will be able to facilitate their needs as an *individual*. This is one of the key factors that causes a person to decide whether or not to be in your structure.

The most important message you can convey to your prospective members is that: They are not just a "number" among the masses. You want everyone who walks through your door to know that you have a grasp of what they need. That you treat each and every one of your students as an "individual", not just a number in relation to your gross profit.

You should not ask them to bring relatives or friends to join your school for any kind of "reward" or monetary compensation, such as a discount on their monthly fees or free gifts. This behavior can be potentially costly in other, more insidious ways. You will potentially send the message to your existing students that you are treating them as a "sales person" for your school and that they are just another "asset" to your "business empire". Also, consider the impact on a "full fee" paying member when they learn that some others are not paying the same full fee, but a discounted one. This can lead to a feeling of unfair treatment and discontent over time.

Also, the simple reality is that discounts equal lesser gross revenues and lower profit margins in your business. *But the concept of "cost" in offering Discounts still remains the same as in all other Investments to grow your business. And you must always aim to reap the maximum return on any of your investments.* Therefore, you should exercise good judgment when constructing each promotional discount offer by going through a "Cost and Benefit" analysis. Read on to gain more perspectives on the subject of ***"Promotional Discount Start-Ups"***.

Follow the "Golden Rules" for all Promotional Discount Start-Ups outlined below:

"Make Money."

The number one rule by which to abide by when rolling out new promotions is to NOT lose money. This must be covered *before* you can think about making money on that same promotion. As the section **"Profit Margin Perspectives"** says, the potential profit in this industry is 60 cents on every dollar grossed.

Let's assume that you are indeed running your operation well. And that you have achieved a fixed cost in your operation of 40 cents on every dollar grossed. That would mean your break even point is to discount up to, but not to exceed, 60% on any one average monthly fee. You will lose money if the discount package offers 61% or more off your average regular monthly fees. Also, if your offer of a promotional special includes merchandise, such as uniforms, shoes, t-shirt, instructional DVD's, etc, you also need to factor in your costs for those items to ensure that you break even.

"Generate immediate and continuous cash flow."

The key to a successful martial arts school is to entice new members and keep them for as long as possible. Offering startup discounts in conjunction with enrolling in a one year membership is a good idea. *But it is imperative that the new member begins paying the regular monthly fee the following month!* This will ensure continuous cash flow. Any longer than a thirty day gap after the initial payment will cause a crimp in your cash flow. Avoid this scenario if possible.

"You need to spend money in order to make money."

Well, yes and no. You must run sensible, appealing promotional packages. Include *free gifts* but keep the price point close to your regular monthly fee. There is nothing wrong with offering what appears to be *deep* discounts as long as you recover the cost of products offered plus any fixed overhead expenses *plus* some additional profit. If there is no profit potential, why do it?

There is nothing wrong with offering what appear to be *deep* discounts as long as you recover the cost of products offered plus fixed overhead expenses *plus* some additional profit.

Experiment with various promotional specials throughout the fiscal year. Keep track of the results of each promotion in order to establish trend and to identify cyclical patterns. This will then help you figure out the proper timing to roll out each promo. Also, you will begin to distinguish your good promos from the not-so-good promos and thereby avoid repeating the not-so-good promos over and over again.

Offer *deep* discount promotions *only* during the slow months to capture whatever foot traffic that does come through your school. Offering *deep* discounts too often will end up cutting too *deep* into your bottom line, and you will see your cash reserves rapidly deplete.

"A catchy phrase with an attractive price point."

You *should* consistently and frequently offer promotional discounts to entice new members to join on a regular basis. Keep your pulse on all the current trends regarding health and fitness. Utilize whatever popular lingo and buzz words that are current and commonly used within your desired market share. Incorporate these same buzz words into your marketing strategies. For example, words and phrases such as *Empowerment*, *Mastery*, *Inner Peace*, *Stress Management*, *Health* and *Vitality* represent goals commonly found within progressive individuals who seek to improve their life situation. Use phrases such as "*Kick Start Your New Year with…*" for those people who will be setting new year's resolutions goals. Structure your promotional specials with words and phrases that resonate with your prospective members and then set a price point that is *too good to refuse.*

"Why offer discounts if they do not make you money?"

> ***It is imperative to institute as your top sales priority, a criteria that each and every new member sign up for a "Long Term Program" in order to be eligible for any discount startup package.***

The motto here must be to *make* money not *lose*

The motto here must be to *make* money not *lose* money. According to national statistics, 50% of all new businesses fail after 6 months. You want your business' "P&L" to be in the positive before you run out of investment capital. Expect that your school will not make money during the initial startup period of 6 months. During the initial phase, your focus must be on recruiting new members to fill your halls. So direct your marketing within your immediate vicinity. Also, you should be running promotional specials in both printed ads and the internet. You must be very clear and concise on what your *COSTS* are for your day-to-day operations, equipment, inventory, and most importantly, your living expenses. The quickest way to close the doors to your school is to *not* keep a fine balance between gross revenue and expenses. Some school owners/operators may think to offer desperate, drastically low cost promotions just to drive traffic into their schools regardless of whether they make money or not. If you do this, your martial arts business will *most definitely* become one of the statistics in the "50% of failed business" on start up.

"The name of the game is to sell your Long Term Programs."

It is imperative to institute, as your top sales priority, a criteria that each and every new member sign up for a "Long Term Program" in order to be eligible for any discount startup package. This tried and true marketing strategy has been implemented by the fitness industry for decades. You should borrow a page from their successful "play book" and modify it to fit your school. Signing new students for short term programs only is not only counter-intuitive but it is also counter productive.

Examples of Promotional Discount Start-Ups that DO NOT Work (part a):

"Try One Month for only $19.95"

Why won't this work? On the surface, it seems like a really catchy and very hard to resist promotion. And it should serve to draw new members into your school, right? After all, $19.95 is such a *cheap* price that practically anyone is capable of spending this amount to try out your school for one month. This part is true. And this is why this may be one of the most common promotional offers seen in schools to entice new members to join. Especially during the initial startup phases of a school, many owners may feel that this campaign will quickly fill their training halls. However, there is a downside. This particular promotional approach puts absolutely no monetary profit into your bottom line. Such a promotional special violates the number one rule about business; "Recoup your costs." This promotion will not even cover your standard overhead costs, let alone any costs of items given away within the promotion. You must consider the aftermath of such promotions before you launch this type of campaign.

> ***In this case you, the school owner, definitively loses even more money in every successful transaction.***

Here are the *hidden costs* of this promotion:

1. New members who enroll in your school under this promotion will be unlikely to want to spend the additional money for their uniforms. So, you will either have new members in your school attending classes without proper uniforms, which creates disruptions in your structure that you do not need. Or, in an effort to prevent such disorder, you will be forced to include *free uniforms* in this promotion. In this case you, the school owner, definitively *loses even more money* in every transaction garnered from this promotion.

2. Expect that most new members generated from such a promotion will stop showing up for classes before the promotional period ends. They may very likely *not* stay for the second month and pay your full rate (remember they came in because of the cheap price). This means that they will effectively default on their contracts with your school.

3. You and your instructors will waste a lot of time teaching such short-term students the basics, effectively taking time away from all your existing, more devoted and productive members in the process. This approach will not support the success of your school.

Wisdom: Your priority should always be to keep your existing students in your school for as long as possible while gaining new ones. *See the earlier chapter **"Maximize Each Student's Potential"*** - Page 60.

Strategy: *Try this instead:*

"$99.00 Start Up Package includes: One Month of Classes, Uniforms, Training Shoes, and DVD." (**See sample promo: *Path To Mastery* - Addendums Two - Page 6)

This promotion is strategically priced close to your average monthly program fees of $100 or higher. With this promotion, you are trying to reach individuals who have a level of disposable income that would allow them to continue to afford your program on a consistent basis into the future and not just the short term. With this promotion, you will be able to recoup your costs of both the products offered *plus* your standard overhead, *and* end up with some profit added to your bottom line. Your new student will appreciate the *free gifts* upon enrolling in your school. And this promotion will also present an opportunity to educate the new student on how to utilize an instructional DVD to aid their studies. This will in turn promote future sales of your Instructional DVD's. The new uniforms and shoes will make the new student feel that they "belong" to an organization from day one *and* you have covered the costs of such uniforms.

Examples of Promotional Discount Start-Ups that DO NOT Work (part b):

"$139.00 Intro Package for Three Months of Classes. Package includes: Uniform, Training Shoes, and DVD."

You should expect that many of these new members will have stopped showing up to your classes before the promotional period ends.

Why not? This seems like a great deal for both the new members and your school. Your school will benefit from the larger than normal individual fee and resulting cash flow immediately, right? Yes and no. There is a *positive Yang* in this situation, true. You get that nice cash flow bump on each sign up. Meanwhile, the new member enjoys a three-months long "payment vacation", not having to pay the regular monthly tuition for the next two months, *plus* the added perks all of the *free gifts* they get. This of course means that they probably will *not* have to purchase anything for at least 90 days. This is the "*Negative Yin*" of this approach. It appears on the surface that your school will immediately generate a higher gross income than average. But, upon closer inspection, the cash flow you should be receiving from each new enrollee will then cease for the next two months. The reality is that you are not generating any profit after the first month on these new enrollees. This break in each new pupil's income stream will actually create problems in your cash flow in the near term. Your student base may result in an increase immediately on paper but this increase will not be reflected in your gross income. Again, this is another promotion that packs some delayed effects.

Here are the *hidden costs* of this promotion:

1. Increased Rate of Default: You should expect that many of these new members will have stopped showing up to your classes before the promotional period ends. Many of them may have no intention of ever paying the normal monthly tuition of $100 or higher. Remember that they have enrolled in your school on the *cheap*. Many of them may default on their contracts after the *Trial Study Course* ends. They may have been only interested in trying it out for the summer for instance and took advantage of the promotion to do just that.

2. Losing Proposition: First, consider that you will need to recover your inventory costs on the uniforms, shoes, and the instructional DVD. This is already a big chunk out of that $139.00 initial payment you receive. Second, divide the residual amount by three, giving you the actual monthly fees collected over three months. Third, subtract the fixed overhead costs over the next three months for that student. More than likely, you will end up losing money in this transaction as illustrated in the sample below (showing a loss of $3.67 per month). Also, both you and your instructors will again spend time in teaching these members without full compensation. See the sample of the cost breakdown below.

Example:	Intro Package	$139.00	
	Inventory Costs	-$ 30.00	
		$109.00 divided by 3 Months=	$36.33
	Fix Costs (*Proposed)		*-$40.00*
	Net Profit/Loss		<$-3.67>

*Proposed Fix Costs are based upon the chapter "**PROFIT MARGIN PERSPECTIVES**".

This promotion would have been a big mistake to implement for any long duration if you are not aware of its negative impact. Prior to rollout of each promotion, you should do a mini "P&L" to be fully informed of the *actual* cost and *profit* inherent in the promotion, such as can be found in the example presented above. This will allow you to be fully prepared for each promotional offer's pros and cons and potential fiscal impact on your operations. This practice will help safeguard against any possible oversight and improve your chance of success.

Strategy: *Try this instead:*

"$249.00 Two Months Intro Package includes: Uniforms, Training Shoes."

(**See sample promotional special - Addendums Two - Page 6)

This is a great Promo for the short-term commitment students. You can see that this promotion is strategically priced to allow you to recoup all your costs of items offered *plus* put a profit to your bottom line that is very near your normal monthly price point *and* allow the enrollee to feel that they are receiving a tremendous value at the same time. You definitely want to foster this perception in your enrollees while *stacking the deck* in your favor fiscally. You should strive for this balance in all promotions offered.

Wisdom: ***Pause for thought before considering "Cashing Out" your next student contract. You may unknowingly be on the path of "Cash Out to Wipe Out"…***

There are companies out there who offer to "buy out" student contracts from martial arts schools for a fee, (oftentimes 10% of the total contract value) and pay the owner a lump sum payment up front for the right to purchase said contract. Imagine receiving a lump sum payment of $2160 when cashing out a 24 month contract valued at $2400, as opposed to collecting $100 monthly payment for 24 months. Sounds tempting, doesn't it? On the surface, it may seem like a good idea in terms of generating a surge in your revenue for that month and it will do exactly that. However, you must understand the long term consequences of "Cashing Out" your student contracts and execute such a strategy wisely and with caution.

First of all, understand that you are effectively borrowing against your "future revenue". Each of your student contracts represents a steady, monthly income to your school for the entire term of the contract. Cashing out a contract means that you are withdrawing that future revenue in one lump sum today. Having such rapid liquidity may cause you to be tempted to spend the "Wind Fall" immediately. And if "improvement spending" is the reason why you have elected to "Cash Out" in the first place, go ahead and do that. This can be a reasonably good investment strategy to employ if you have to upgrade broken down or nonfunctional equipment, fund a necessary renovation or expansion or perhaps make a special purchase for your upcoming event or seminar, etc. Such expenditures will in theory generate either immediate or near term returns. BUT, and this is a big but, it is an especially bad idea to cash out contracts in order to pay back rent or delinquent bills. You will be committing the egregious error of "Digging a deeper hole" for yourself. Read on for further elaboration.

Service Fees = Reduced Profit Margin.

There is a service fee attached to Cashing Out. Therefore, you are getting paid *less* than the actual contract value immediately. Let us suppose that the service fee is 10% for cashing out each contract. The cost for this service *must* be subtracted from your bottom line immediately. If your profit margin is at 60%, or 60 cents on every dollar grossed, cashing out a contract will give you a reduced net of 50%, or 50 cents on every dollar grossed. Not too much of a price to pay, right? Not so.

You will get paid less for teaching more students. If the cost for cashing out a contract is 10%, it would mean that you have to teach one additional student per every ten students in your classroom to make up the difference. The *extra* student in your classroom will take your time from other students, causing all students to receive *less* than desired attention needed in their learning process. The quality of your service provided to your students suffers. To rectify this, you will have to hire assistants to balance the teaching load. This will in turn *cost* you labor dollars. If you are not extremely careful, you will end up being out of pocket far more over the long run than you would be if you *did not* cash out contracts.

You will get paid less for teaching more students. If the cost for cashing out a contract is 10%, it would mean that you have to teach one additional student per every ten students in your classroom to make up the difference.

The other extreme danger of the cash out syndrome is that it is entirely too easy to cultivate a mindset wherein *you think you have more than you have!*

All successful businesses rely on a steady and consistent cash flow in order to properly balance their revenues and financial obligations. A sudden "Windfall" can give many *new* business managers a false sense of confidence with regard to the actual performance of their business. If you were to cash out *all* contracts as soon as they are signed and then use the proceeds to pay your monthly overhead, you will doom yourself to running out of money in short order. The reality of the martial arts business is that no one can anticipate how many future new enrollees may enter your school in any given month, or month after month. The desire to cash out can be a common mistake that any new owner/operator *may* be likely to make. But we must remember that this practice at its best can cause a serious crimp in your cash flow during a down trend. Or, at it's worst, deplete your capital entirely before the next contract year begins and students are due to sign new contracts again. If you make use of the cash out methodology too often, you will have to wait until that time rolls around again before you can start collecting tuitions again. So, be advised… Execute with caution if you *must*. But only if you must.

Tuition/Enrollment Strategies

First of all, understand that having a "Student Contract" is a necessity in building a successful school. If you are still not using contracts, you are a dinosaur on the brink of extinction. And it is only a matter of time. One big economic downturn will push you over the brink.

When we first began teaching, back in 1978, we didn't use contracts. We had an altruistic belief that "if you have something of real value, the students will stay of their own accord". Well, this is nice in theory and it may indeed be true for a small percentage of students. But it is also true that small percentage will not pay your bills, let alone fuel your school to success. All operating health club chains have understood this for decades. That's one big reason they *are* a *chain*.

Now I know, many of you reading this may say that if they don't want to train you don't want them there, right? Wrong. You do want them there. It is a fact of life in our industry that you will service 100% of your population so that you may truly teach less than half of them.

Look at it this way.. A student cannot profit from their studies if they are not there, right? If you do not keep them enrolled and try to encourage them to be there, they have no possibility of gaining anything from their studies, right? Therefore by keeping them enrolled on a contract you are, at a minimum, leaving the door open for them to attend. No matter how infrequently they actually DO attend, you are at least giving them the option and opportunity to benefit.

Allowing them to quit after only a month is benefiting nobody. We all know it takes longer than that to start to see any form of progress in the martial arts. So, instead of making it easy for them to quit, make it harder via a contract and then attempt to motivate them to actually stay focused in your programs. Oftentimes their reasons for not attending can be traced to a feeling of being lost and this can be alleviated by giving them a little private tutoring in the basics of any respective class they are attending.

How to sell the Contract

Essentially, the key to this is found in the above few paragraphs. It is easy to sell a contract if you believe and truly understand that students are indeed making a commitment to themselves via a contract so that they may allow themselves the time to devote a focused effort to the studies. Martial arts skills are not formed in a day. And often the initial few months are spent "wading" through the unfamiliar waters of the basics in a completely foreign environment. It takes a few months for the "fog" to begin to lift and for the student to actually begin to see and feel the effects the training is having on their mind and body.

Therefore, selling the contract becomes a simple effort of making them understand these facts. And this also gives you an excellent opportunity to establish yourself as the helpful and caring teacher you know yourself to be by telling them you are there for them at any time to help them. After all, they are indeed there because they want to better themselves and you are indeed there to help them accomplish that goal. These efforts take time and not allowing enough time to really try is not helping anyone. So a contract is really a win/win situation for all concerned.

Tiered Enrollment Structure

We highly recommend a tiered enrollment structure. See the example below:

Example is fictitious. Determine what your own market can bear and charge accordingly

Term Option #1:

(includes 1 private lesson per month)

Duration of Term: 18 Months

Monthly Payment: $150.00

Term Option #2:

(includes 1 private lesson per month)

Duration of Term: 12 Months

Monthly Payment: $160.00

Term Option #3:

(includes 1 private lesson per month)

Duration of Term: 6 Months

Monthly Payment: $180.00

Term Option #4:

(includes 2 private lessons per month)

Duration of Term: 18 Months

Monthly Payment: $180.00

Term Option #5:

(includes 2 private lessons per month)

Duration of Term: 12 Months

Monthly Payment: $190.00

Term Option #6:

(includes 2 private lessons per month)

Duration of Term: 6 Months

Monthly Payment: $210.00

Signups will tend toward the lowest price per month. But *you win* no matter which price point they choose and this is an important thing to remember.

Those prospects that are timid often choose the lowest time duration because of the uncertainty they feel about their involvement. Hence they pay you more per month, either initially or while they are there for their first term. If you have faith in your school and its programs, a prospect choosing this option is just fine for you because that student will likely upgrade when their term is up and choose one of your long term programs. In the meantime, you actually reap a higher reward for their timidity.

Such a student would be in class as an extra body, more work for you, no matter what price point they chose and so them choosing to pay more is simply a bonus to you. Meanwhile, you are able to *"sell"* that prospect by letting them limit their time commitment to the study via a shorter agreement making the "contract pill" easier for them to swallow in the process.

The more self assured prospect will tend to choose a longer term right away, providing long term security for your cash flow and revenue stream. These students literally become the life blood of your school assuring that you can continue to grow your population base. And you are able to *"sell"* to their boldness on a long term contract by enticing them with a lower fee per month to "reward" the commitment they show to themselves in their pursuits.

A tiered structure is simply easier to sell. It allows you to be more flexible in your sales pitch and it allows you to be far more nimble in dealing with individual mind sets in your prospects. This flexibility combined with a more nimble sales approach will make it easier for you to sell any and all contracts to your prospects and provide a solid growth potential for your school.

Sign up Packages

Let's face it. We live in a society where people have come to expect something free to earn their business. So you simply must indulge that need in your prospects.

A well planned sign up package can both satisfy the need for "free" in a prospect as well as earn you revenue in the process. Below is an example of a promotional package you can use as a template for construction of your own.

Note: More can be found at the back of this manual. And more discussion on sign up packages can be found in the chapter titled: **"Maximizing Your Income Potential"** - on Page 45 of this book.

Example is fictitious. Determine what your own market can bear and charge accordingly

PATH TO MASTERY STUDIES *(For the more committed, long term student.)*

A $364.00 Instructional package for **ONLY $189.00**

Your *Path To Mastery Course Includes all you need to get started right...*

• 1 month of Group Classes. (no limit attendance, come every day)	normally $ 140
• 1 Introductory Private Lesson.	normally $ 60
• Association Fee Included. (registers you internationally)	normally $ 45
• 1 Free Uniform. (top & bottom included)	normally $ 45
• 1 Complete set of School Patches plus training shoes.	normally $ 26
• 1 Free School T-Shirt (12 Animal Design; 6 colors)	normally $ 18
• 1 Beginner's Video Tape or DVD. (Featuring all required forms & Applications.)	normally $ 30

All of this is only $*189.00* when you sign up as a new student under any of our regular student plans. $*189.00 is all you pay for this entire one month Path To Mastery course.* Your regular plan will commence immediately after your Path to Mastery course ends. What a great beginning!

Charge What You're Worth

Many in the martial arts industry feel that there should be an attempt to undercut competition with regards to our business pricing. But this perspective is not only wrong, it will cause your school to struggle, sputter and perhaps die. This mindset must be broken to be successful.

Look at it this way. A doctor spends 8 years in school and another 3 to 5 years in their residency afterward. They are very well compensated for their acquired knowledge base, and should be, earning a six figure income or more annually. How much time have YOU spent in acquiring your current knowledge base and skill set within the martial arts? You ARE AN accomplished professional. And as such, you should expect the same kind of substantial fiscal reward for your professional expertise.

Endeavor to excel at what you do in your school. Allow all your prospective members to see and realize the superior quality contained in your programs, your devoted student base, your school and you. People like to associate with quality. Show them that they are on the path to quality in their education in every way and you will earn another loyal student.

Remember, perception of value is everything. Find out who charges the most in your area and set your price points higher than theirs. We are not saying be greedy or unrealistic. Just set your price points higher by a reasonable margin. Then take appropriate steps to exceed on all levels, and in all areas, with regard to the services your competition can offer and to deliver a value that is far above their best.

The temptation to charge a lower price point than your competition comes from the mentality that you may be "missing out" on many prospects that could otherwise afford your services. But be aware that lower price points tend to attract a "lower quality" client and , as a result, can be more headache than reward. Be advised that using lower price points may bring you a much larger than desired cross section of the following three types of individuals:

The violinist - This is the student whom is always late with his or her monthly fee and always

has a ready made excuse for "getting it to you next week". They will always refuse to be on an auto debit structure because they "just don't trust that" enough. And as a result, you will spend much of your valuable time trying to get this person to pay you. Meanwhile, they will merrily attend classes and take time away from paying students all the while forestalling their payment to you.

The antagonist - This person comes into your school with a chip the size of lake Michigan on

their shoulder and then proceeds to not only antagonize other students while in your school, but to also question anything and everything you do and then question the answers to their questions. Not that questions are bad, but this type of person often derails the class structure to such a degree that the entire class loses focus and walks away frustrated. Let this happen too often and this individual can cost you many good students. After all, your students are there to learn from you, not to hear this person talk about whatever inane perspectives they may hold.

The Neanderthal - This individual has zero tolerance for anything that he or she has deemed as

"non-traditional". They may gripe about the fact that you use air conditioning, claiming that it is not Spartan enough for training "for real". Or they may talk about the "good old days" when things were harder than they are now, even though they themselves were never part of the good old days. The kind of rhetoric that flows from their mouths is like a slow acting poison. If unchecked, it can seep slowly but surely through your entire student population causing unrest and low morale. Which will, in turn, lead to mass dropouts. Avoidance of such individuals is generally the best strategy.

Billing Strategies

There are numerous billing strategies you can employ in today's martial arts school marketplace with regards to individual or specific "tools". But all such tools effectively fall into two broad categories. And you must weigh the pros and cons carefully based on factors such as increased overhead, time involved in processing and your own individual personality. Yes, you read it right, your own personality.

In-house vs. "Outhouse"

There are numerous billing companies out there which are for hire nowadays. And they are, all of them, only too willing to do this nasty chore for you as long as you give them their share of your profits in the process.

Many of today's professional martial arts associations strongly advocate this tactic claiming that it is just too much trouble for you to "wear the hat of a bill collector" in addition to a school owner's numerous other responsibilities. They seem to feel that the activity of bill collecting may take way too much of your precious time and focus off your main goals of providing stellar service within stellar programs. Interestingly, if you look very closely, often these same professional associations own a billing company and give their members "discounts" for using the billing service through them. Hmm.

We would be the first to agree that the following idea does take some time in setup. But given today's computer technologies, once you do have it set up it takes practically no time at all.

With a few form letters, a database management program, and an ability to merge the two it takes only minutes a month to generate a couple of hundred billing letters if necessary. Besides which, you should have most of your student body converted to an auto pay method anyway, right? So you shouldn't have all that many letters outstanding each month to worry about. *And* all of these necessary functions can be conveniently found in a single office program like say *Microsoft Word.*

If you don't want to stuff the envelopes yourself, pay a "postnet" or some other such company to do it for you. It will cost you FAR less than the up to 10% of each student's monthly tuition that billing companies often charge. Simply put, billing "out of house" is equivalent to the old classic *Outhouse,* or, down the toilet for your hard earned money.

In addition, if you are truly worried about your personality not being able to handle the down on his/her luck, just lost my job, weeping student syndrome occasionally. Well, there is good news for you. "In house billing" can be "sold" as out of house billing, keeping the additional 10% within your bottom line profit AND your students will never be the wiser. This is how:

Step 1) Set up a dba (doing business as) "XYZ Billing Co."

Step 2) Register and file the company separately from your school entity.

Step 3) Enlist a P.O. Box for the fictitious company where it can receive mail from anyone wishing to contact the "billing company".

Step 4) Use computer generated form letters to handle the nasty business of sending bills when you need to do so.

And just like that, you have not only saved 10% of your profit margin, you have also insulated yourself from the occasional sob story of a student whom is having trouble paying his or her tuition and wants you to bear the burden of same.

You now have in place an always available, ready made escape by simply being able to say to such a student, "I'm sorry, but I can't do anything about that. You'll have to talk to the billing company. Here is the mailing address." And then provide the P.O. Box address.

Need More help with Your Operations?
Hire Us as Outside Consultants.

MASTER MIKE PATTERSON:

He is a *Highly Successful Owner/Operator* for Adult's Only Chinese Martial Arts Schools. His first school, established 1978, was voted one of the top ten in the nation. He was one of the first to utilize the world wide web for promotion of the martial arts and he continues to be cutting edge in all aspects of martial arts promotion. He is the *Founder* of the Innovative Martial Program R.S.P.C.T. (realistically structured progressive combat training). He has acted as *Publisher* for an International Martial Arts Magazine. He has been and still is a *Producer* of a full line of Internationally recognized Instructional Videos since 1988.

email us through: www.hsing-i.com

To Franchise or Not to Franchise

In the distant past, when younger, we once attended a business seminar on martial arts school franchising. The presenter made the point of saying that you need to "keep your people happy" and "provide a place for them to grow" in the same business they came up within. He advocated franchising your school to your more hard working instructors and providing an avenue for them to grow that way.

The problem with this scenario is simply that this sort of individual usually does not possess the financial means to "pay their own way" in the process of opening a school. Nor do they have the credit or means to acquire a loan to cover any of the associated expenses. Which means that you will be footing the bill for such an operation in hopes that your student "investment" will pay off in the future.

There are many ways that such an arrangement can be set up. And it is not our purpose to go into the various methods here. Our purpose of discussion is to only let you know this can be, and is, risky. And it will cost you much in terms of time, money and aggravation. So be sure this is what you want before you go down that road at all.

Such individuals, because you have taken all the risk on their behalf, tend to have no "pride in ownership" of the school they run. They also can come to hold the mindset that they are not easily replaced. This is why, if you do decide to franchise in any way, it is imperative that you have both a systems driven curriculum (see **"SETTING UP YOUR CURRICULUM"**), AND a valid and productive instructorship program in place to provide a pool of potential should you encounter a need to suddenly change management of any give satellite.

One well run school can net $250K

Yes. One well run school can net, not gross, NET $250,000 per year. And it can do this with a minimal staff of two, plus a few well trained assistants.

The benefits of one quality school are numerous and multi-faceted, much like a polished gem. You have far better quality control. Students all want to be trained by the Master or Masters anyway. And in a single school environment you can give them exactly what they want. They want your constant presence, even if only in a mostly supervisory position. And your constant presence will serve to promote a strong perception of value and morale, helping you to establish long lasting rapport with your student body and keep them involved. Of course, you will want to hone your people skills. Nobody stays with a grouch for 10 years or more of their lives.

In a single large school, you have far less headaches in regards to management. There are only one set of books to keep. One pro shop to stock. One curriculum to organize and promote. One set of restrooms to keep clean, etc. etc. etc.

You also have less personnel issues to deal with. There are no paid instructors (the same of which can become your future competition). Having no paid instructors means a reduction in your liabilities. Also, your insurance premiums will stay lower. There are no payroll taxes to deduct and no employee health insurance to provide, all of which will translate directly to your bottom profit line. There are no Labor Laws or Labor Board to contend with to make your life more complicated than it should be.

In a single school you will have far better cost control and gross potential/up-sells because you have a vested interest to be successful as owner. Employees seldom have such vested interest and, in fact, often give you only what they have determined is a fair day's work for the money you pay them. And this perspective, if held, often translates to lost students due to a severe lack of quality service.

As opposed to Multiple - read inferior, Satellite Schools

Multiple locations will require you to "grow your own" personnel within your structure since you want to promote your style in these satellite schools. The very necessity of doing this will mean that these same protégés will become familiar with ALL of your business strategies. Having this knowledge will put them in a strong position to compete against you. In addition to this, up until the time that they can truly go it alone, they often turn in a substandard work ethic, on your dime. This can end up costing you time, money and possibly many good students in the process.

IF you choose to play the franchise game, at the end of the day the best case scenario is that you stay on good terms with this person and that they buy that same location from you in terms of franchise. This will thereby relieve you of most of the corresponding headaches that go along with that other location. BUT, this is not always how things work out.

In fact, oftentimes such individuals do not follow through and you end up being saddled with the additional location along with your main one, now plagued with personnel problems. And trying to run it effectively can cost you dearly. PLUS, this person you have trained is now trying to become your competitor down the block. So think carefully and protect, protect, protect yourself in the process if you choose to go this way.

Consult an attorney about what protections are afforded to you in your state and then have your attorney advise you on the best strategies for your circumstance.

If your wish is simply to expand your own "brand" of martial arts in a given territory, you might consider an alternative route. Those of your *Instructorship Program* graduates who are serious about opening their own school could indeed be allowed to open their own school. But set it up so that these individuals are meeting their own financial obligations without you providing the financial backing. This way, their success or failure will not affect your personal finances and you still spread your respective discipline across a given area.

Consult an attorney about what protections are afforded to you in your state and then have your attorney advise you on the best strategies for your circumstance. DO NOT make the mistake in thinking that franchising is easy, no matter what form it takes. It is not. And you DO need a great deal of good advice to set up such a structure.

Profit Margin Perspectives

What is "Profit Margin"? Profit Margin is that percentage of every dollar grossed which will end up in your bank account at the end of each month *after* paying all of your expenses. What is the current "Profit Margin" in your martial arts business? Let's make this simple to understand.

Ask yourself this question: How much of the $100 payment your student make to you each month actually ends up in your bank account after you pay all of overhead expense on each individual student?

Ask yourself this question: How much of the $100 payment each of your students make to you each month actually ends up in your bank account after you pay all of overhead expense on each individual student? Take that $100 and pay all of your operating expense for that student.

Let's suppose that at the end of each fiscal month, the residual amount is $25 per student that ends up going into your bank account. This would mean the "Profit Margin" in your business is 25%, or your school generates a net profit of 25 cents on every dollar grossed. If this is the case, you can do much better.

Main Street Vs. Kung Fu Street - *Net profit of 10 cents vs. 60 cents on every dollar grossed*

Let's begin with the obvious. The number one fundamental of *any and all* business operations is to generate a profit. But this is very likely an area of education that was not covered in your studies to become a professional martial arts instructor/owner operator.

You must understand that there are unique differences, in terms of operating costs, when you are comparing the martial arts business on "Kung Fu Street" to the conventional businesses on "Main Street". You must learn how to capitalize on the advantage of the lower operating cost in a martial arts business. This will in turn help maximize the potential of your profit margin.

You can see the variables in terms of the operating costs of a business on "Main Street" verses a business on "Kung Fu Street" in the examples of Profit and Loss (P&L) statements shown on the page immediately following:

Example A–	"Main Street" businesses
	"Projected Profit & Loss/P&L" sample statement breakdown:
Projected Annual Gross Revenue: $420000	
Essential/Necessary costs in daily operation that are Manageable:	
Product Cost	Typically 25-35% of the gross revenue
Labor Cost	Typically 15-20% of the gross revenue
Total Cost	at approximately 40-55% of the gross revenue
Fixed Cost Categories:	
Advertising	
Insurance	
Rent	
Utilities	
Taxes, Fees, Misc., etc.	
Total Cost	at approximately 40-45% of the gross revenue
Total Operating Cost:	at approximately 90-95% of the gross revenue
Projected Annual Profit: 5-10%	*$21000 to $42000 or 5 cents to 10 cents on every dollar grossed*

The fictitious "Main Street" businesses referenced as "*Example A*" would be your regular retail shops found in most strip malls and shopping centers. Shops such as: apparel, shoes, toys, books, corporate chain stores, fast food restaurants, etc. These types of businesses necessitate employees to *sell* or *serve* the products, post production, to their customers. The operating costs of these two categories ranges between 40% to 55% of the gross revenue in such businesses. See *Example A* in "Essential/Necessary costs in daily operation that are Manageable".

The "Fixed Cost Categories" section applies to all industries across the board, as illustrated in both examples found in this section. These costs will be determined by the individual type of business. But such costs will be fixed and non fluctuating, except for cost of living increases, once the business is established.

The final line indicates the *"Projected Annual Profit", which for the "Main Street" businesses ranges between 5% to 10%, or a profit margin of 5 to 10 cents on every dollar grossed.*

Example B– "Kung Fu Street" businesses

"Projected Profit & Loss/P&L" sample statement breakdown:

Projected Annual Gross Revenue: $420000

Essential/Necessary costs in daily operation that are Manageable:

Product Cost	0% (It cost you zero, *You are The Product*)
Labor Cost	0% (It cost you zero, *You are the Service Provider to your students*)
Total Cost	0% of the gross revenue or "ZERO" dollar

Fixed Cost Categories:

Advertising

Insurance

Inventory

Rent

Utilities

Taxes, Fees, Misc., etc.

Total Cost	at approximately 40-45% of the gross revenue
Total Operating Cost:	at approximately 40-45% of the gross revenue
Projected Annual Profit: 55-60%	*$231000 to $252000 or 55 cents to 60 cents on every dollar grossed*

In Contrast, the "Kung Fu Street" business referenced in *Example B* is the martial arts business. The lower operating cost in this type of business can be easily defined in both the categories of "Product" and "Labor" cost. This remarkable disparity is simply because a professional martial arts instructor/owner operator *IS THE "PRODUCT"*, therefore *zero product cost* is incurred. He or she is also the "Service Provider" to their students because they are teaching all of the classes. Therefore *zero labor cost* is incurred as well (assuming you have no salaried instructors on staff and that you and your partner do all the teaching). This means that the "Kung Fu Street" business *does not* have to bear the cost of "*product*" and/or "*labor*". This significantly reduces the operating cost by up to 40% to 55% of the gross revenue.

This is fundamental reason that enables the "Kung Fu Street" type business to generate such a lucrative projected profit margin of approximately 60% of the gross revenue, or 60 cents on every dollar grossed!

You can clearly see the stark difference in terms of *"Projected Annual Profit"* in the prior examples. Does your martial arts business generate a profit margin of approximately 60 cents on every dollar grossed? If not, and you are reading this manual, you have now seen the real potential. You should be eager to optimize the overall performance of your martial arts business to realize that same potential.

There are two basic key business principles in achieving a higher profit margin; 1) *increase revenue* and 2) *reduce expenditures*. Focusing on both of these sides of business will ensure more profit goes into your bank account at the end of each month.

Does your martial arts school generate enough profit to cover your expenses for both personal and business needs? If you have to maintain another job just to make the ends meet at the end of each month, then likely not.

Key Principle #1: Increase Revenue

A martial arts business generates its revenue from several sources; tuition, private lessons, tests, seminars/workshops, retail sales and annual membership fees. It is essential that the "business manager" in you should track these categories separately giving you a definitive view of your business performance. Tracking things separately and meticulously will enable you to detect the shift in trends in each category of your gross revenue. The information gleaned will allow you to adjust your business strategy to ensure the continuous growth of your revenue/cash flow. Tracking things meticulously allows you to "be more nimble" in making changes quickly, and as needed. And being nimble is key to succeed.

Does your martial arts school generate enough profit to cover your expenses for both personal and business needs? If you have to maintain another job just to make the ends meet at the end of each month, then likely not.

Are you maximizing your sources of revenue in every category? If so, you should be able to become a full-time professional martial arts instructor with a substantial monthly income.

You should realize the massive potential gross revenue of a martial arts business. Once you have made this comparison between "Main Street" and "Kung Fu Street", you will be ready to develop marketing strategies for the implementation of broad sweeping changes to develop a lucrative business model that is capable of out-performing most conventional businesses.

Facility Operations

The proposed scenario in this chapter is based on a perceived ideal school operation of two full time instructors in a start-up *or* in an already existing school facility.

The optimal study program proposed in this scenario includes two private lessons per student, per month, in addition to regular group classes. In a "45 hour work week" two instructors can handle a total of 150 to 200 students.

You could also add more hours to your work week and accommodate a student base of 250 with two instructors. For example, adding 5.5 hours per week per instructor will allow you and your partner to accommodate an additional 50 students in your student base.

Let's look at the examples below in different potential scenarios of fictitious schools run by two instructors with active members of 150 and 200:

Sample A is based on "Two Instructors" with a work week of 45 hours per instructor. This proposed optimal study program offers *two private lessons* per student, per month, plus group classes. The "Projected Profit Margin" is at 60% or 60 cents on every dollar grossed.

Projected Monthly Fees		/ Active Members	Projected Annual Gross Revenue		Projected Profit Margin	Projected Annual Profit
$140	x	150 = $21000 per month x 12 months =	$252000	x	60%	=$151200
$140	x	200 = $28000 per month x 12 months =	$336000	x	60%	=$201600
$150	x	150 = $22500 per month x 12 months =	$270000	x	60%	=$162000
$150	x	200 = $30000 per month x 12 months =	$360000	x	60%	=$216000
$175	x	150 = $26250 per month x 12 months =	$315000	x	60%	=$189000
$175	x	200 = $35000 per month x 12 months =	$420000	x	60%	=$252000
$200	x	150 = $30000 per month x 12 months =	$360000	x	60%	=$216000
$200	x	200 = $40000 per month x 12 months =	$480000	x	60%	=$288000

If your student base were to grow to 300 students or more, two instructors would only be able to offer a *less* than ideal study program. This would include only one private lesson per student per month in addition to group classes. In addition, each instructor will be required to have a schedule that exceeds 45 hours per week. You will also likely need to hire a staff to run certain aspects of your daily operations as your student body grows to beyond the maximum capacity of a "two person" operation. You must staff your facility according to the size of your student body so that you do not compromise service quality. This will ensure the continuous growth of your school.

Sample B is based on "Two Instructors" with a work week of 45 hours per instructor. This *less than ideal* study program offers *only one private lesson* per student per month plus group classes. The "Projected Profit Margin" is at 60% or 60 cents on every dollar grossed.

Projected Monthly Fees		/ Active Members	Projected Annual Gross Revenue		Projected Profit Margin	Projected Annual Profit
$140	x	250 = $35000 per month x 12 months =	$420000	x	60%	=$252000
$140	x	300 = $42000 per month x 12 months =	$504000	x	60%	=$302400
$150	x	250 = $37500 per month x 12 months =	$450000	x	60%	=$270000
$150	x	300 = $45000 per month x 12 months =	$540000	x	60%	=$324000
$175	x	250 = $43750 per month x 12 months =	$525000	x	60%	=$315000
$175	x	300 = $52500 per month x 12 months =	$630000	x	60%	=$378000
$200	x	250 = $50000 per month x 12 months =	$600000	x	60%	=$360000
$200	x	300 = $60000 per month x 12 months =	$720000	x	60%	=$432000

***"Monthly Fees" include tuition, annual membership fee, testing fees, seminars, retail purchase, etc. See the section on **Maximize Each Student's Gross Potential** - Page 61, for complete breakdown of projected revenue generated from these same categories.

Summary:

You *can also* implement **Sample B** with a proposed optimal study program that will offer two private lessons per student, per month, plus group classes. But the reality would be that you would need "three instructors" with a work week of 45 hours per instructor. This will still yield the same projected profit margin of 60% or 60 cents on every dollar grossed. But you will now be dividing the profit by "three" instructors. Or, you could consider a single, salaried instructor as opposed to a third "partner".

How many active students do you have in your school at the present time of this reading? Does your school exceed the maximum occupancy of your facility? It is highly likely that your school does not perform to its maximum potential compared to the two samples shown above.

Okay. I'm Not There. What Do I Do Now?

First, you should zero in on whether or not you have maximized all "Sources of Revenue" in all categories in your school. Make a comparison of what each of your students pays on an average to the proposed monthly fees in the samples.

Let's suppose that each of your students pays $80 per month. Now compare that to the lowest fees proposed in the samples of $140 per student. You can already see a disparity of $60 per student that you are potentially not earning. Consider that you may be undercharging, no matter what your area market competition is doing.

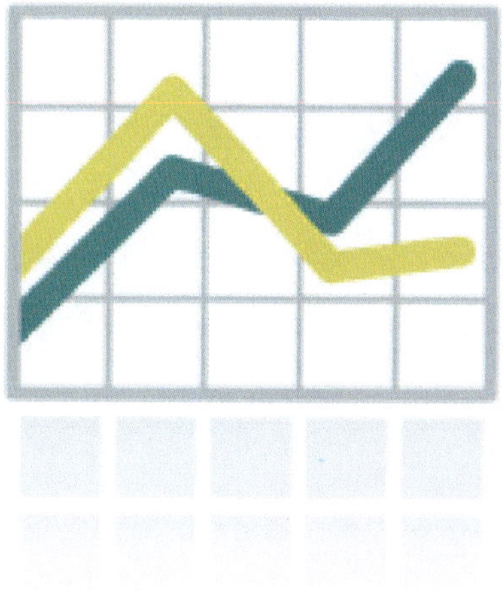

Now analyze your actual records in "Sources of Revenue" and identify the categories that are not 100% optimized. Is your "mark up" on product sales reasonable? Are you charging a reasonable fee for testing? Are you holding regular seminars? Look at all these categories carefully.

You will now have a clear and concise direction to go in making the necessary adjustments to improve the categories that are less than ideal. There is also *probably* room to grow in terms of your student base (meaning the maximum occupancy in your facility).

Second, focus your aim in student retention. See the chapter **"STUDENT RETENTION STRATEGIES"** for suggestions. Achieving and maintaining lesser numbers in the "drop-outs" and "non-renewal" categories translates to a growing student base, assuming your school gains new enrollees on a regular basis. It should, or you need to scrutinize your marketing.

Third, develop marketing strategies that target increased student enrollment. See the chapter **"ESTABLISHING A MARKETING PLAN"**. Be ready to promote new and improved study programs and watch the new enrollees eagerly move toward becoming a part of your structure. This will ensure that your school begins to generate the revenue you have always deserved.

Now that you have addressed principle #1, increasing revenues, you next need to learn specific methods and the importance of how to reduce expenditures. Successfully implementing this knowledge will allow you to achieve and realize your target of a lucrative profit margin of 60% of the gross revenue, or 60 cents on every dollar grossed.

Key Principle #2: Reduce Expenditures

To get a hold on your expenditures, the first thing you should be doing is generating an accurate Profit & Loss (P&L) statement of your business at the end of each fiscal month. Once you have implemented this "Monthly P&L" practice in managing your business, you will begin to develop a crystallized perspective of the actual performance of your martial arts business on a monthly basis.

Secondly, having this data from your monthly P&L, you should now create a "*budget*". The data in your monthly P&L will allow you to establish what your *true* "Fiscal Budget" is based on the actual known costs/expenses. This will then allow you to have a realistic guideline, to which you can adhere, to optimize your cash flow.

But wait. Before you start moving forward, you must first construct an accurate business budget prior to beginning the process of "trimming" your expenditures. Otherwise, you will be making the mistake of "entering the battle field wearing a blindfold". The results *will* be catastrophic and the potential damage is likely not reversible.

Making A Budget

Understand that there are some basic ideas you must grasp before you can create a budget that is both *real* and *attainable*. It becomes imperative that the records kept in your monthly accounting reflect your *actual* expenditures in order to create a *real budget.*

It is equally important that you document irregular events that cause unique, "one-time charges". You should subtract the "one-time charges" from the total expenditures when deriving the final budget. Otherwise, you will have an "inflated budget". And to adhere to this

It is common to find the categories of "Fixed Cost", "Fixed Cost with Fluctuation" and "Non-Fixed Cost" contained within a standard monthly and/or annual "Profit and Loss" (P&L) statement with regard to a conventional business.

inflated budget in the following months or year will result in a reduced profit margin.

Each "one-time charge" should be kept in its own separate category in terms of the amount spent and time/day it was spent, especially the types of expenses that incur "every so often". Tracking these types of expenditures separately will allow you to plan better, and allocate funds for these "one-time charges" when they become necessary. It will crimp your cash flow if too many of these "one-time charges" hit your account in a short span of time.

The ideal scenario is to space out these events so that you will have the cash flow to support these costs and not affect your daily operations.

Costs Categories

It is common to find the categories of "Fixed Cost", "Fixed Cost with Fluctuation" and also "Non-Fixed Cost" contained within a standard monthly and/or annual "Profit and Loss" (P&L) statement with regard to a conventional business.

Within these categories, there are what are called "Recurring" costs (those costs which incur monthly and annually regardless), and "Non-Recurring" costs or "One-Time Charges" (those costs which do *not* incur monthly or annually but rather once in multiple years).

It is essential that you and your business partner understand the definitions of each category in your "P&L" statement and how each category affects your students, prospective members and daily operations. Read on in the following pages for full explanations of these terminologies and their associated items:

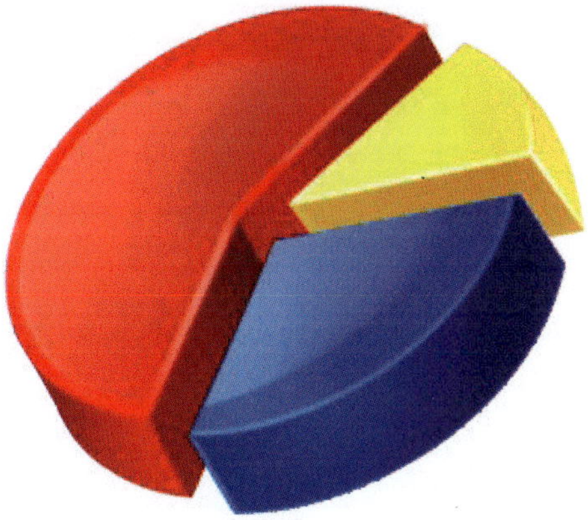

A *"Recurring Fixed Cost"* would be defined as those expenses that incur monthly, at a fixed amount, *without fluctuation*. This usually applies to such things as building leases, service contracts, annual advertising contracts, etc. The *Recurring Fixed Cost* category includes:

Rent; including "Cam Charges" or "Triple Net Charges"

Advertising; in phone books or other contracted advertising

Janitorial, Repair & Maintenance Contracts

Liability Insurance

Detailed descriptions for the *"Recurring Fixed Cost"* category:

Rent: This, of course, is a standard cost in most businesses unless you own the space where you conduct your business operation. Your monthly rent or lease payment will be structured based on the existing lease between you and your landlord. The lease will usually include "Cam" or "Triple Net" charges if you lease a room or building in any commercial strip mall or a shopping center. The "Cam" or "Triple Net" charges are fees that landlords impose on the tenants for regular maintenance and upkeep of the business center where you lease to run your business. For more details on this topic refer to the chapter on **"NEGOTIATING A LEASE"** regarding "Cam Charges". The rent for your martial arts business *recurs monthly* and stays at a *fixed* amount until the lease expires.

Advertising: The local phone companies will typically publish printed phonebooks annually in most cities. Publishers of the local phonebooks will then require a one year contract for running your printed ad. You will pay a *fixed* amount and it *recurs monthly* until your contract expires. Any other contractual agreement in advertising should also be in this category.

Janitorial, Repair & Maintenance: This is a standard cost in *all* businesses that service the public. It is *necessary* and it *recurs monthly.* The cost for any and all janitorial supplies used to maintain your restrooms, training equipment, devices and the entire facility should be included in this category. Additionally, you may choose to outsource this area and hire a contractual Janitorial and/or Repair and Maintenance Service to maintain your facility. Both the supplies and the outside services should be in this category.

Liability Insurance: This is a standard cost in *most* businesses, especially if they perform services to the public. Liability Insurance Providers for the martial arts industry usually require a one-year contract for their underwriting policies. The amount will be *fixed* and it *recurs monthly, quarterly or annually.*

A *"Recurring Fixed Cost with Fluctuation"* entails those costs and charges that *fluctuate* with the gross revenue. And such expenditures will depend on the specifics of the category. These costs also *incur monthly* and include:

Advertising; that is not contracted, but elective

Credit Cards Fees

Inventory Costs

Office Expenses

Utility Costs

Miscellaneous

Detailed descriptions for the *"Recurring Fixed Cost with Fluctuation"* category:

Advertising: This is exclusive to all advertisement to both prospective members and your existing student body on a month-to-month basis. These expenditures incur from handouts and pamphlets, mailings, flyers, lead boxes, etc. The expenses in this category *recur monthly* but they *fluctuate* based on need and elective choice.

Credit Card Fees: Each time a credit card is swiped on your credit card terminal, you are subject to a transaction fee. The transaction fees *recur monthly* and *fluctuate* based on the volume and proceeds from your total credit card sales. Also, if you do not own the credit card terminal, you will be on a lease to rent the equipment. The monthly leasing fee for your credit card terminal *recurs monthly* and is usually set by the equipment leasing company at a *fixed* amount.

Inventory: Items you purchase to stock your retail shop and/or "free gifts" you hand out as part of any promotional offers belong in this category. These costs *recur monthly* and they *fluctuate* based on what is sold and/or given away.

Office Expenses: Paper, pens, pencils, ink cartridges for your printer, software for your computer, etc. belong in this category. The *monthly fluctuation* is based on usage.

Utilities: Costs in cooling, heating and telephone should be recorded in this category. The *fluctuation* is based on actual usage and *recurs monthly*.

Miscellaneous: There are always some "odds and ends" expenses that do not belong in the conventional categories. For example: decorative wall art for the purpose of beautification, a water fountain in the changing room, a student comment box, magazine subscriptions, etc. just to name a few. The *fluctuation* is based on discretion or necessity and it *incurs regularly*.

A *"Non-Recurring Cost"* entails those expenses that DO NOT recur every month on any consistent basis. These events usually relate to those costs related to maintaining or upgrading your facility that are either a "one-time charge" or once every few years or so. You *must* keep track of such *necessary* spending and then budget for both immediate and future planning. It is imperative to leave room in your budget for such expenses so that you may avoid the pitfall of not being able to perform the repair or improvement in a timely manner. The *"Non-Recurring Cost"* category will include:

Equipment Purchases

Improvements spending

Office, Library expenses

Repair/Maintenance expenses

Detailed descriptions for the *"Non-Recurring Fixed Cost"* category:

Equipment Purchases – Heavy bags, Sparring mats, training devices, etc. are items that need to be repaired, replaced or upgraded over time. These will incur once every few years or less.

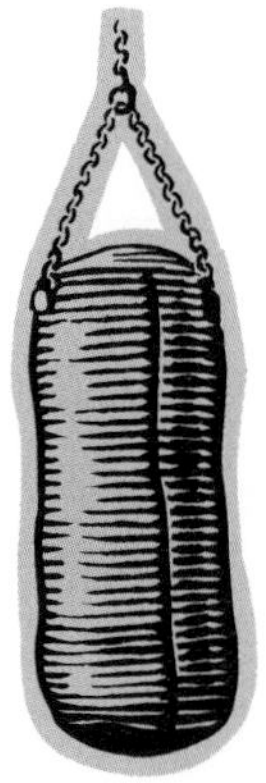

Improvements– New exterior sign, new furniture in the lobby or visitors lounge, restroom or changing room fixtures, retail display cases, etc. are items that need to be repaired, replaced or upgraded over time. Such expenses will incur once every few years or less.

Office, Library– New computer, printer, furniture, telephone, DVD player,

TV, books, charts, instructional DVD's, etc. are items that need to be replaced or upgraded over time. They will incur once every few years or less. You will likely (and should) also increase the collection of study aide materials in the Students' Library periodically.

Repair/Maintenance– New carpet, flooring, interior paint, light fixtures, etc. are necessary expenses that incur once every few years or less. As such, they belong in this category.

The *"Non-Fixed Cost"* category entails those expenses that are mostly "one-time" charges and are "Non- Recurring". This category includes:

Improvements

Legal Services

Marketing costs

Renovations

Detailed descriptions for the *"Non-Fixed Cost"* category:

Improvements: It is important to replace and update old and/or worn out amenities such as training equipment, your store front, mirrors in training halls, student locker rooms, etc.

Legal Services: It is very wise to seek professional legal assistance when needed. This is a necessary expense and incurs one-time or infrequently depending on the issue. See the chapter **"STICK TO WHAT YOU ARE GOOD AT"** for more on this perspective.

Marketing: One-time radio or occasional advertisement to promote new ad campaigns or during startup or expansion phases.

Renovations: These expenses often incur during the startup phase, or an expansion. Expenses incur in renovating or expanding training halls, restrooms, changing rooms, etc. and are usually one-time expenses during startup, expansion or relocation only.

Creating Monthly P&L Statements

Now that you have become acquainted with the definitions of all types of expenses relating to your business operations, get ready to start making your own regular monthly "P&L" statements. Begin with properly documenting all expenses, in their appropriate categories, of your monthly "Profit and Loss" or "P&L" statements in order to create a business budget. Following are some simple steps to create your business budget. You have to start somewhere. See below for a sample of a twelve month "P&L" statement:

Sample 12 Month "P&L" statement of a typical martial arts school: (the figures are fictitious)

January		February		March	
Gross Revenue:	$6384	Gross Revenue:	$5802	Gross Revenue:	$5619
Rent:	$2000	Rent:	$2000	Rent:	$2000
Utility:	$ 235	Utility:	$ 220	Utility:	$ 165
Advertising:	$ 568	Advertising:	$ 515	Advertising:	$ 482
Insurance:	$ 100	Insurance:	$ 100	Insurance:	$ 100
Inventory:	$ 220	Inventory:	$ 205	Inventory:	$ 187
Janitorial:	$ 75	Janitorial:	$ 44	Janitorial:	$ 52
Office Supply:	$ 115	Office Supply:	$ 87	Office Supply:	$ 96
Misc:	$ 217	Misc:	$ 176	Misc:	$ 77
Total Expenses	$3530	Total Expenses	$3347	Total Expenses	$3159
Profit	$2854	Profit	$2455	Profit	$2460

April		May		June	
Gross Revenue:	$5343	Gross Revenue:	$5701	Gross Revenue:	$5597
Rent:	$2000	Rent:	$2000	Rent:	$2000
Utility:	$ 157	Utility:	$ 144	Utility:	$ 187
Advertising:	$ 500	Advertising:	$ 451	Advertising:	$ 462
Insurance:	$ 100	Insurance:	$ 100	Insurance:	$ 100
Inventory:	$ 164	Inventory:	$ 197	Inventory:	$ 172
Janitorial:	$ 83	Janitorial:	$ 63	Janitorial:	$ 91
Office Supply:	$ 113	Office Supply:	$ 68	Office Supply:	$ 80
Misc:	$ 165	Misc:	$ 506	Misc:	$ 69
Total Expenses	$3282	Total Expenses	$3529	Total Expenses	$3161
Profit	$2061	Profit	$2172	Profit	$2436

(continued next page)

(continued from previous page)

July		August		September	
Gross Revenue:	$5085	Gross Revenue:	$4963	Gross Revenue:	$5384
Rent:	$2000	Rent:	$2000	Rent:	$2000
Utility:	$ 237	Utility:	$ 396	Utility:	$ 245
Advertising:	$ 457	Advertising:	$ 436	Advertising:	$ 493
Insurance:	$ 100	Insurance:	$ 100	Insurance:	$ 100
Inventory:	$ 94	Inventory:	$ 128	Inventory:	$ 155
Janitorial:	$ 59	Janitorial:	$ 44	Janitorial:	$ 68
Office Supply:	$ 42	Office Supply:	$ 83	Office Supply:	$ 106
Misc:	$ 29	Misc:	$ 54	Misc:	$ 147
Total Expenses	$3018	Total Expenses	$3341	Total Expenses	$3314
Profit	$2067	Profit	$1722	Profit	$2070

October		November		December	
Gross Revenue:	$5605	Gross Revenue:	$5710	Gross Revenue:	$5425
Rent:	$2000	Rent:	$2000	Rent:	$2000
Utility:	$ 209	Utility:	$ 187	Utility:	$ 219
Advertising:	$ 428	Advertising:	$ 475	Advertising:	$ 520
Insurance:	$ 100	Insurance:	$ 100	Insurance:	$ 100
Inventory:	$ 225	Inventory:	$ 243	Inventory:	$ 185
Janitorial:	$ 85	Janitorial:	$ 72	Janitorial:	$ 69
Office Supply:	$ 94	Office Supply:	$ 76	Office Supply:	$ 115
Misc:	$ 66	Misc:	$ 82	Misc:	$ 179
Total Expenses	$3207	Total Expenses	$3235	Total Expenses	$3387
Profit	$2398	Profit	$2475	Profit	$2038

You would like to compile twelve months of "P&L" Statements. One year of "P&L" statement will show cyclical patterns throughout the year. To start, you should have at least three to six months of "P&L" Statements if you are just beginning to implement the strategy.

Take the total of each category and divide by the number of months i.e. 12 (months), 6 (months) or 3 (months).

The figures resulting in each category will become your *monthly budget* in said category. You will notice the figures in the "Fixed" categories remain consistent throughout the 12 month period. And that the "Non-Fixed" categories fluctuate on a monthly basis.

The practice of generating a monthly P&L statement will present an immediate opportunity for you to effectively analyze the performance of your martial arts business once you have the fiscal budget in place for all categories. Remember, the key to maintaining optimal performance is to be *"in the moment"* and act *nimbly.*

Simple truth; finding out that you are over-budget in the middle of the year is better than you finding out at the end of the year. It will be too late for you to take any action to balance out any overages if they are only detected at the end of the year. You want to become aware of such things developing early on in that same year. You do not want to spend *ahead* of your allotted amount and you would rather rein in your spending spree *as it happens*.

See samples of a revised budget from 3 months to 6 months and 6 months to 12 months below:

Sample A: Aggregate of 3 months "P&L" and mathematical calculation to derive the monthly budget:

January –March Total			Monthly Budget	
Gross Revenue:	$17805	Divide by 3	Gross Revenue:	$5935
Rent:	$ 6000		Rent:	$2000
Utility:	$ 620		Utility:	$ 207
Advertising:	$ 1565		Advertising:	$ 522
Insurance:	$ 300		Insurance:	$ 100
Inventory:	$ 612		Inventory:	$ 204
Janitorial:	$ 171		Janitorial:	$ 57
Office Supply:	$ 298		Office Supply:	$ 99
Misc:	$ 470		Misc:	$ 157
Total Expenses	$10036		Total Expenses	$3345
Profit	$ 7769		Profit	$2590

Sample B: Aggregate of 6 months "P&L" and mathematical calculation to derive the revised monthly budget:

January –June Total			Monthly Budget	
Gross Revenue:	$34446	Divide by 6	Gross Revenue:	$5741
Rent:	$12000		Rent:	$2000
Utility:	$ 1108		Utility:	$ 185
Advertising:	$ 2978		Advertising:	$ 496
Insurance:	$ 600		Insurance:	$ 100
Inventory:	$ 1145		Inventory:	$ 191
Janitorial:	$ 408		Janitorial:	$ 68
Office Supply:	$ 559		Office Supply:	$ 93
Misc:	$ 1210		Misc:	$ 202
Total Expenses	$20008		Total Expenses	$3335
Profit	$14438		Profit	$2406

Sample C: Aggregate of 12 months "P&L" and mathematical calculation to derive the revised monthly budget:

January –December Total			Monthly Budget	
Gross Revenue:	$66618	Divide by 12	Gross Revenue:	$5552
Rent:	$24000		Rent:	$2000
Utility:	$ 2601		Utility:	$ 217
Advertising:	$ 5787		Advertising:	$ 482
Insurance:	$ 1200		Insurance:	$ 100
Inventory:	$ 2175		Inventory:	$ 181
Janitorial:	$ 805		Janitorial:	$ 67
Office Supply:	$ 1075		Office Supply:	$ 90
Misc:	$ 1767		Misc:	$ 147
Total Expenses	$39410		Total Expenses	$3284
Profit	$27208		Profit	$2267

You should revise your budget as you accumulate more "P&L" Statements to establish your actual *fiscal budget*, especially if you started with only a partial year of data or any number that is less than twelve months. A partial year of "P&L" Statements *will not* give you an accurate fiscal budget because of the absence of cyclical factors. Therefore, it is imperative to complete a twelve month compilation worth of "P&L" Statements before finalizing your *fiscal budget*.

(Sample of the projected fiscal budget for the upcoming new year, using the current year end figures plus annual inflationary factor of 3%:)

January –December Total			Monthly Budget	
Gross Revenue:	$66618	Multiply by 1.03 then Divide by 12	Gross Revenue:	$5718
Rent: $24000			Rent:	$2060
Utility:	$ 2601		Utility:	$ 223
Advertising:	$ 5787		Advertising:	$ 497
Insurance:	$ 1200		Insurance:	$ 103
Inventory:	$ 2175		Inventory:	$ 187
Janitorial:	$ 805		Janitorial:	$ 69
Office Supply:	$ 1075		Office Supply:	$ 92
Misc:	$ 1767		Misc:	$ 152
Total Expenses	$39410		Total Expenses	$3383
Profit	$27208		Profit	$2335

Now that you have created your fiscal business budget, get ready to begin monitoring the actual performance of your martial arts business. This process will help you to see the gaps between the actual expenditures and your proposed budget in all categories. Also, any discrepancy in actual gross revenue and your projected gross revenue will become obvious. This information will allow you to be nimble in adjusting your marketing campaign when you notice a "dip" in your gross revenue. You should learn to exercise your analytical skills by doing regular "P&L Analysis" well before executing any "cuts" in your budget.

P&L Analysis

What is "P&L Analysis"? P&L Analysis is a practice which every good business manager applies to scrutinize their Monthly P&L Statements. This process takes place after the end of each fiscal month and it is crucial to do this again at the end of each year as well. You will gain valuable knowledge from analyzing the history of your business's performance and then be able to apply the wisdom gained to both the present time and to the future.

A routine practice of P&L Analysis will help ensure the optimal performance of your business operations which translates to long term prosperity. You may also occasionally discover certain charges or costs that should *not* be there. Sometimes, such charges in your P&L are caused by outdated and/or antiquated arrangements which you have unknowingly inherited from the last tenants or from your landlord. Here are some incidents that might be seen in your analysis:

Over or Under Budget in Inventory

<u>Loss of Inventory</u>: It is common that every so often the costs in Inventory Purchase may spike for reasons such as seminars, promotions, sales, etc. You should investigate to identify the source of the rising cost. For example, you may discover that one boxed shipment filled with uniforms was misplaced and you have repeated the same purchase order. Or, perhaps some shipment was delivered and left outside your front door when your school was closed, though you never actually received it because it was stolen. This could cause a shortage in your retail business effectively causing "out of stock" items until the replacement order is re-delivered. These types of mistakes are caused by human errors, which can and should be avoided. You should designate an area for deliveries, and/or persons who are to manage inventories. Assign a less hectic weekday and specify delivery *only* during your hours of operation if possible.

Mismanagement in Ordering: The assigned person(s) who manage the inventory should be trained to verify accurate "inventory on-hand" prior to placing a new order. Also, it will be helpful to keep accurate sales records of items sold, as well as to track the sizes of uniforms, shoes, etc. This will help prevent overage of any "less in-demand items" such as a size two uniforms or a size forty six shoe for example. These less in-demand items will just sit on your shelves for months without turning any profit. By the same token, you do not want to be sold out of the "high-demand items" such as size five uniforms, popular size shoes, T-Shirts, etc. And meticulous inventory tracking will assist in this process.

Increase of Wholesale Cost: Your suppliers *do* and *will* raise the cost in products they sell to you for variety of reasons; change of supplier sources, increased cost in raw material, inflation, new owners seeking higher profit margins, etc. Keep a close eye on your invoices to spot such changes. Find out how you can get the products at a lower price, perhaps by ordering in larger quantities or by finding new suppliers who have access to lower cost products. Business is about making money. Do not feel obligated to certain vendors just because you have had a business relationship with them that stems way back from a decade ago.

Vendors Cost Increase

Be meticulous in scrutinizing charges from vendors who are setup for monthly or quarterly recurring billing for their services, such as janitorial services or A/C maintenance companies. You *must* examine these categories carefully each month to spot any increases that are over your budget. Re-negotiate a lower fee or find new vendors with more competitive rates.

Power Spikes

Over Budget in Utility Cost: This actually happened in one of our schools. For years we had been paying roughly $35 per month for a rather small suite (less than 200 square feet). It was such a nominal amount that no one ever took a second look at it. Years later, during a power crunch, the local power company decided to implement a substantial increase of cost because there was a major shortage of the power supply in our city. We were being charged more than triple the amount of our normal bill and it persisted for months without any adjustment forthcoming. Upon further investigation, assisted by the representatives from the power company, a scenario of "power siphon" was revealed by both the landlord and the recent new tenant adjacent to us.

The cost amount to thousands over the years. This siphon would never have been detected if there had not been the substantial rate increase imposed by the power company. We would not have paid increased costs for years if we had been more meticulous about our P&L Analysis.

Antiquated, Outdated Contracts

There was a time while working as a general manager for a "Fortune 500 Company" that monthly P&L Analysis was a part of my duty. The very first time I analyzed the P&L, going line-by-line in my job, I discovered two charges for "utilities" occurring on a monthly basis. This was unusual for a "common, non business person" like me such as I was in those days. And I wondered; "How come my store gets billed twice for utility consumption when my own residence only gets one bill per month?" I posed this question first to the regional office and received no explanation. Then, the query was forwarded to the corporate accounting department and also to several other officers to get to the bottom of why this dual charge was there. This caused the company I was working for to begin an extensive investigation of the issue, which included review of old contracts that had been established and were in place decades ago. Through investigation, they discovered that some old and outdated contracts had been carried forward to the present time simply because it had been overlooked all those years, given that the charge seemed "nominal". When all was said and done, that *nominal* over-charge amounted to millions of dollars to the company when totaled together over the course of decades across all of the stores in that particular region.

The simple moral of such incidents should be "No minor overage is too minor. Investigate as soon as you spot one. It could save you a lot of money in the long run."

Rent & Cam Charges Increase

Rent Increase: Landlords will typically raise your rent incrementally within your lease term based on the projected cost of living increase. Or, it is also possible that you could face an over-night skyrocket increase if you are not on a lease. Try to re-negotiate for a longer lease term or offer concessions such as paying for some level of tenant upgrade or general repair of your facility to maintain a lower percentage of rent increase. Do a cost and benefit analysis to ensure your deal is equitable for both you and your landlord.

Cam Charges Increase: This is an area that some new owner/operator may overlook. The reason is simply that "Cam Charges" are commonly based on the square footage of your facility. And, as such, often referred to as pennies per square foot. Some may think that "25 cents per square foot" is not that much to pay. Right? Think again. Let's suppose your school occupies 2000 square foot of space. Now multiply that by 25 cents. You will end up with an increase of $500 per month in addition to your monthly rent. Let's go out further to one year. That is $6000 annually. Do your math before agreeing to any proposal presented by your landlord. When in doubt, seek expert help. See the section on "**STICK TO WHAT YOU ARE GOOD AT**".

Training Your Front Person

One of the most often overlooked aspects of running a martial arts school is the training of the front person. Yeah, I know.. Anyone can do it, right? And of course your senior students are all good candidates because they know a great deal about the arts you teach and can therefore "wow" any and all visiting prospective students with their knowledge and skill. Wrong! Keep such people on the training floor where they belong.

A good front person is indeed knowledgeable about your school's programs, but need not be all that familiar with the arts you teach. That would ideally be the domain of the head instructor who will come by to briefly chat with the prospect later on in the evening. This will be done AFTER the prospect has watched the scheduled class, had their walkthrough of the premises and been told about the numerous and beneficial programs your school has to offer. Or, in the alternative, your friendly and well trained front person could hand them a concise pamphlet with such information presented inside, which they can then peruse while watching a class.

Personality, access and knowledge count.

Face it, friendly sells. If the front person is a likeable person, your school has already made that ever important "good first impression". Make sure the person you choose is of such character. They should come across as easy to talk to and intelligent. They should also be helpful and courteous. It should be clear to the prospect that the front person is there to ease their stay, for as long as they may stay. Whether that be just for the night, or as a permanent member, the front person must make the prospect feel welcome and at home.

A pleasant appearance never hurt anyone.

This is not bias. This is marketing. From our experience, we highly recommend that the front person be a "she" rather than a "he". Females tend to perform much better in this role than do their male counterparts. A female is more able to relate well to the inevitably predominant male population of prospects coming to your school without that pesky male ego getting in the way. And, she is also better suited to relate to the female population of prospects and/or existing students coming to join your school. A female front person can develop almost instant rapport with another female and make her feel comfortable.

It is also a well known scientific fact that people tend to respond better to those individuals that they find pleasant to look at over those they don't. This is, again, not a bias. This is science and such decisions should be governed by fact only.

Hire someone to serve as your front person that fills the above criterias and it will pay multiple dividends once they are well trained.

Where to find and how to train your front person.

A good, and likely nearby, "gene" pool from which to find and select your front person would be a local community college or university. College students are always looking for part time jobs so that they can earn some money to work their way through school or pay for books or weekend fun. Try placing an ad in such places, even on campus bulletin boards if funding is tight. Make sure the ad says "will train" in the context.

Remember, this person will become the "Face" of your school. This will be the person whom a prospect first deals with initially and they must be trained in a cohesive and systemized manner. If you are pulling from colleges, you will end up having to train a new front person regularly. Any college student will tend to "move on" fairly quickly unless you give them a substantial reason to stay. And even then they will still go when school is finished and it is time for them to embark upon their chosen career. This is not a bad thing as long as your training method is systemized.

The number one priority is to make sure they are training in, and familiar with, your programs. Yes, I said training in your programs. Get them involved in your school even if you do it as part of a compensatory "perks" package which lets them train for free. This is so very important so that this person will develop a sense of enthusiasm for what your school can offer the public. If you do not do this, imagine this scenario:

Prospect - "Do you train here?"

Front Person - "Oh no, I do pilates at the gym. Besides, the uniforms here make me look fat."

Sound funny? Believe us when we say *it can and does happen.* And you definitely do not want such things rolling out of your front person's mouth. You want them to say things like; "Oh yes! I train here and all the instructors are WONDERFUL!!"

Dealing with Walk-Ins

This part, unfortunately, is not so easy to teach. You can role play this aspect to a degree, and you should. But prospects standing live, in person, can easily throw curve balls at your front person and it takes awhile for most front people to learn to adjust on the fly. A simple example would be that a prospect may ask the front person "Do you spar here?" Such a question is a loaded question, and a new front person will not recognize it as such. They may say "yes, we do spar" or "no, we don't spar" thinking this is the true answer (whichever is true in your school) and not understand that the question is dual edged. Selecting the wrong response can send the prospect away without any further exploration into your school or its programs simply because they sought the *opposite* of the front person's given response.

Initially you should stick to the basics. Instruct your chosen front person that they are to be *certain* to make appointments for *every prospect* to both view classes *and* meet the Master Instructors without exception or failure. Until your front person has better skills, most questions regarding the specifics of any study need to be handled by the instructors on site.

The front person should be relegated to understanding the *sign up programs only*! Meaning all those materials regarding sales, that you have so thoughtfully printed up and laid out for all of your prospects to read through, while waiting to speak to you personally.

If the prospect attempts to question the front person on anything regarding the specifics of study, they should be told something like "that is a very good question and it would be best answered by the chief instructor." They should NOT risk speaking out of turn and, as a result of their lack of knowledge or understanding, risk alienating the prospect.

Dealing with Phone Inquiries

We highly advise the process of scripting responses for FAQ's (frequently asked questions). And then do role play both via phone and in person to acclimate your front person to such scenarios until they handle the phone effortlessly and do not sound "canned" when dealing with prospects. It is very important that your front person's "phone presence" sounds natural to the prospect as this will come across as confident. We developed many such role play evaluation systems over the years to great effect. See the following page for a sample of a good P.T.E. (phone training evaluation form):

Phone Training Evaluation (P.T.E.)

When you call, keep it short, do what is necessary to be believable but do not tie the front person up on the phone too long. Avoid taking them away from their real duties. Award up to the point value listed in parentheses behind the blank. If the value is one, award either one or zero. Use only whole numbers. Award points based on accuracy (1 pt. ea. item), presentation and believability (1 ea). Choose one of the lead-in questions, score the answer you receive, then score the rest of the sheet based on follow up responses from the person answering the phone. There are 58 points possible. We will start the standard at 35 and try to improve…

Possible lead-in questions:(use some creativity)

Option #1: How much do you charge?

Answer #1: It varies, depending on the type of program you select. ____(3)

#1a: First, let me tell you a little bit about the School… ____(3)

Option #2: Could you tell me about your program?

Answer #2: Sure, let me begin with a brief explanation of our curriculum ____(3)

#2a: We have a great deal of variety and flexibility. ____(3)

Option #3: What style do you teach?

Answer #3: We teach four complete systems of Chinese Martial Arts, we also teach Chinese Medicine, Chi Kung, Meditation, Philosophy and Chinese Language. What are you looking for? ____(3)

#3a: What you are looking for relates to........(fill in the blank) which we do teach.... (elaborate) ____(3)

Caller Comments: ____

Follow Up Analysis:

Point #1: Our curriculum is extensive, so... (Ask caller to get pen and paper) ____(4)

A: We offer unlimited attendance; can come to every class every day… ____(4)

1. We have classes weekly in four different styles of Chinese Martial Arts. For Adults; also teach; Chi Kung; Meditation; Philosophy; Chinese Medicine; and Chinese Language… ____(11)

2. First Group Class starts each day at noon, last one at 8:30 p.m.; Private lessons can be done almost any time. We closed Sundays. ____(5)

3. We also have three training halls here at the School; so during peak hours we have up to three different classes to choose from during any given hour; that means there are classes available to all levels of students several times each day… ____(5)

Point #2: (Move through this quickly and authoritatively)

A: Prices range from $110 to $135 per month depending on the length of time you wish to invest in your studies… we also offer a discount startup package for you to consider. ____(4)

1. For example; if you want to make a commitment to your goals, we give you a lower rate per month; if you want to go month to month with no commitment, you will pay a bit more; in either case, there is no limit on attendance… ____(5)

2. We do not require any kind of down payment, you pay only your first months fee to get started. The only other fee connected to your starting classes here is a $45.00 yearly membership fee to the (insert your Assoc. here)… ____(4)

3. The best thing for you to do would be to stop by and pick up our free information packet… ____(3)

Point #3: Try to schedule appointment. ____(4)

A: Name ____(1) B: Phone # ____(2) C: Previous experience ____(1)

Total Points = ______

Comments: ____

You should run such training scenarios with your personnel frequently and consistently until your front person scores high enough on a regular basis to give both you and that individual the confidence necessary to fulfill their role.

They will be much more comfortable, as will you, knowing they have done everything possible to be prepared for their role. There is absolutely nothing worse than a poorly prepared front person in relation to your operations. This person will make your school's first impression on any prospect that visits you. And you must understand that your front person's performance will directly and strongly influence how your prospect feels while being in your school for the very first time.

It is your job to ensure that your front person has the right aptitude, personality and training to do the job you are hiring them to do for you. Do not take this task lightly or approach it in a haphazard manner. Doing so can cost you dearly.

Overcoming Objections

Understand that objections must never be taken personally. Objections are simply a part of the process and they must be taken in stride and dealt with properly. Some prospective members will need to be handled differently than others. And some prospects will require some additional persuasion to feel comfortable joining your program.

Remember, you do not know their back stories. They may have been burned before in another school or program. Or, they may be somewhat intimidated by the idea of studying martial arts, even though they are indeed interested. It is your job to find out what the real objection is and deal with it in a positive manner.

It is often true that they are unsure that they *want to spend* the money even though it is true that *they can afford* to spend the money if they so choose. You must remain calm, helpful, alert and concerned while working through this process with them.

Following are some examples of likely objections which you or your front personnel may encounter in dealing with prospective enrollees:

I can't afford this.

Translation; you have not yet convinced this person that you offer a *value* that is equal or greater than the fee you charge. Most people can afford it, but they have to choose to spend.

Answer: Well, we have numerous programs to choose from and I am quite sure we can find something to fit your budget. Please remember that study is an investment in the overall quality of your life and personal security. Let me help you choose a program that is right for you.

Why are you more expensive than .... ?

Answer: You are familiar with the saying, "you get what you pay for"? Well, it is as true in the martial arts industry as it is anywhere else. You simply cannot compare our school to any others. We offer a vastly superior program and our instructors are some of the most qualified in the ... city, country, world (state the truth). In this school, I can personally guarantee that you will receive a value for the money you spend in tuition that far exceeds your expectations. No other school can make that same claim.

Why should I study at your school over any other?

Answer: Our school and its programs speak for themselves. You have already watched a class and I know you have now seen that our students and their skills speak for and represent the institution and the ability of the instructors here to help you realize your goals. We have a long track record in this city of properly serving our students and providing the best education available in the martial arts, bar none. You will find no better school anywhere in the city.

I need to consult my spouse.

Translation; You have again not convinced this person to spend.

Answer: Wouldn't your wife/husband want you to make a decision that is in your best interests in terms of your health, your personal security and your well being? Why don't we schedule another appointment right now and bring him or her along to talk to me in person? I'm sure that once we all sit down together, we can come up with a program that will fit you and your budget perfectly. (this strategy may also lead to signing up the spouse)

I'm not sure this will work with my schedule.

Translation; You have not yet convinced the prospect. They are looking for reasons to leave your office.

Answer: We have classes every day of the week, except Sundays, at all hours of the day. I'm sure that we can find a program that will work seamlessly with your schedule, no matter what obstacles you may be facing. Please tell me what your concerns are currently with scheduling and I am sure I can work with you to find a solution.

I need to think it over.

Translation; It's probably a money issue, but the prospect does not want to state this outright as they are embarrassed to admit they cannot afford your school or your programs.

Answer: I see that you stated your primary goals of training are.. (list these from their filled out application). Is that true? When they say yes, ask; Is it that you think this school cannot provide what you need to reach those goals? When they say no, ask; Is it a money issue? Then work with them to try and find a solution.

I don't have enough money with me to sign up today.

Answer: That's okay. What kind of a deposit can you leave with me today? We can work out a payment plan for the remainder of your balance and still get you started right away. (this is why it is also very helpful to accept all major credit cards)

I would prefer to watch a class before I sign up.

If you have set up your sales and marketing strategies correctly and in accordance with earlier recommendations found in the chapter "Establishing A Marketing Plan", the prospect should have already watched a class that is relevant to their interests before you are even having this "closing" conversation with them. But, if this is a "walk-in" and they are seemingly interested, you should without hesitation agree and schedule an appointment for them on the spot. Remember to get their full name and contact phone number. Put it into your schedule on the spot and make a follow up reminder via phone call on the day of, or night before, the scheduled appointment.

What if I only come one day a week?

Answer: Perhaps you may start with one day a week. But once you become involved and see the benefits firsthand of what our programs can do for you, I am quite sure you will want to come to your designated classes more often. Besides which, I strongly feel that with only once per week attendance, we will be unable to assist you properly in achieving your goals. (If you happen to have a once a week program, sign them up under it and then attempt to upgrade them later. If you do not have such a program, find the one that is closest to that but offers greater value for their money and "up sell" them on the spot to that program).

Always remember that objections are not to be taken as personal criticism of your school, your programs, your staff or yourself. They are simply stepping stones to bringing the new prospect into your existing student body as a productive member. Address their concerns and questions with kindness and concern. Display a genuine willingness to work with them to overcome any issues they may be facing and you will not only earn a new student but you will also, in the process of truly trying to assist them, earn a loyal student as well.

Student Retention Strategies - Stop the revolving door dead in its track

The biggest issue in the martial arts business becomes one of student retention. Some students are going to quit no matter what techniques you employ. This is just a fact that must be accepted. Lives may change. Students move because of job change. Crises come and go in their lives, etc.

These circumstances cannot be avoided and oftentimes they are not something you can help. But, many students often quit for reasons that CAN BE helped or even prevented. And this is something you absolutely MUST be mindful of all turns to grow your school successfully.

Perception of Value

The most important thing overall is to understand that you need to create a perception of value in the experiences your students have while within your school walls. This perception of value must be carefully crafted and woven into the fabric of every aspect of your school functions, curriculum and institution. Failure to do so will result in your students losing faith in what they are learning and drop away from your curriculum. Remember; "It is always easier to retain an existing student/client/customer than it is to gain a new one".

> *In the beginning phases of training, new students often experience a feeling of a lack of personal attention in assisting them to integrate into your structure.*

Everyone expects to feel a little "green" but not in an overwhelming way. And let's face it. It is somewhat daunting to those inexperienced in the martial arts to be confronted with all these new movements, patterns, customs, etc. that go along with studying.

You can alleviate this to a great degree by designating an assigned personal assistant for each new pupil. This person is assigned to stay with them during their first few classes and assist them in overall understanding so that they may more seamlessly integrate into their curricular level. This is generally a much better approach than having an "introductory class" or two where they still will be lumped in with several other students. It also will cost you less in terms of manpower hours as the older student assigned is already there for class anyway. AND as a side benefit, for this purpose you may be able to either use or convert one of your students into your instructorship programs.

As the pupil continues in your program, a feeling of a lack of attention to progress made is often manifest in their minds.

It is imperative that you utilize "positive reinforcement" when dealing with all of your pupils. Attempt to always say something positive to each and every student in each class which they attend regarding their effort. Always find something good to say before you criticize in an effort to correct. For example: If a student is practicing a side kick on the floor and not doing it to standard, don't walk up and immediately say; "That's wrong. Do it this way." Instead, say; "You're really starting to get the flow of that technique. Try changing this (give your advice) and I think you may even see better results than you are already experiencing.

Each student must leave every session with something they can "hold" onto.

You must be sure that each student gained something from their participation in each and every class they attend. One way that you can help to make sure this is true is to construct and utilize "lesson plans" so that you keep the flow of material you teach both fresh and consistent.

Watch the eyes and body language of your pupils as they work on a new technique, form or paired drill. If they are not understanding, or worse, getting bored with the task at hand then you must immediately take action to clarify by further demonstration. Or, change it up entirely to keep them feeling like they are moving forward. It is imperative that they leave each class with a feeling of true progress and increased insight into what they are learning.

Don't expect the modern student to "eat bitter" like in the good "old" days.

Yes, I know. We all did and why shouldn't they, right? I mean, after all, this method worked to produce us didn't it? These are different times. And most of your students are not in your school to become the "badass" that you are today. Believe it or not, most of them are there primarily for intellectual pursuit of the martial arts and the delusion of being a badass. Yes, I said delusion. Now, not all fall into this category mind you, which is why you need to have a diverse program to accommodate the ones who truly DO want to follow in your footsteps and master the art in question. But MOST of them certainly DO NOT want that, even if they say they do. It is a mistake to not understand this.

"Good Old Days Mentality" = Dinosaur = Extinction

Nowadays, comfort sells. A modern student lives and works in a comfortable environment. They will expect no less from the environment that they now train within. Modern gyms provide all amenities a person could ask for as part of their membership. People are used to this level of comfort and service.

In contrast, most martial arts schools cannot provide amenities like showers, a sauna, a pool, etc. Therefore, you must make every other available effort to make their experience not only rewarding but comfortable in the process. Never forget where the adult professional comes from and cater to creature comforts wherever possible.

For example: Turn on the A/C during all business hours. Do not try to save a few bucks and then watch your student body leave because of your oversight.

Don't do parking lot classes under any circumstances. If you have that much overflow, it's time to consider other options like adding new classes or expanding your facility if at all feasible. A typical adult professional is not going to be happy doing their training in a parking lot for very long. They may not express their dissatisfaction directly to you. But they will soon walk if the circumstance does not change and change quickly. Besides which, it is likely that your liability insurance does not extend to the public parking lot in any case.

Building Brand Loyalty

Understand that progressing students are happy to remain complacent forever within your school. And these same people will become your biggest source of new enrollees if you give them that feeling of deep satisfaction they crave.

That "personal touch" goes a very long way in establishing this kind of loyalty. Greet them by name when you see them. Make a point to remember small things that they tell you about their lives. Ask after their daughter's recent scholarship, that recent business trip they took or about that wedding they recently attended.

You must display a genuine interest in each of them as the individual they are and must not treat them as "just another student". Your school can become their sanctuary, their rock in the storms of their lives if you allow it to happen. Make it a reality for them.

Track attendance carefully and systematically.

Make phone calls to absentees when they fail to show up to classes within their normal pattern and inquire as to what may be wrong. And, most importantly, if there is anything you can do to help. Do not send a card or an email. A phone call is far more personal.

Keep track of your well attended classes.

But also track your lesser attended classes and then make the necessary adjustments to keep your student body happy and fulfilled. Tell the entire student body via posting a notice on your bulletin board as to why the change has been made. Make it very clear that you hear them and empathize with their needs and that you are willing to try and accommodate those needs.

Keep a "Student Roster."

And then perform meticulous maintenance regarding each student's vital stats in your school. Things like their current rank, the materials needed to achieve their next rank or sub-rank as well as their projected test date based on past performance are all very important. Let one slip through the cracks in this regard and that same student will feel slighted and begin to think you do not care about them. And then they will leave.

Prioritize your operations.

Existing students must *always* take priority over prospective students. A prospective student in your lobby must *always* take priority over a prospect you are talking with on the phone. You must show all people you interact with at all times that the people who are loyal to your school are treated with respect and consideration. And that all other business takes a back seat when one of your own has a need to be fulfilled. This behavior, if consistent, will build a loyal following that cannot be shaken, no matter what the circumstances. They will choose to stay because they love what is represented by you and your school. They will choose to stay because they feel a part of, and cared for, by your school.

Existing students must always take priority over prospective students.

Cultivate the Five Virtues.

The tradition of the five virtues says Honor, Integrity, Respect, Loyalty and Humility. Everyone equates martial arts schools with such virtues. And they expect that both the school and all the personnel within them, will be exemplary in character.

You must always be mindful that you are being watched and judged by your student base at every turn, for everything you do and everything you say. You must always "Lead by Example" and you must endeavor to be an "impeccable warrior". You must be beyond reproach, therefore having nothing to be attacked. Behave professionally and act in accordance with such innate professionalism at all times. Stay within your character as the teacher you are and do not ever deviate for any reason.

Special Programs and Classes are an effective retention tool.

Have you ever wanted to belong to an exclusive club of any kind? Odds are you have. What made you want to be a member? Was it the fact that this club was not for everyone? Why go to Harvard Law as opposed to any other law school? Why shop at Tiffany's as opposed to any other jewelry store?

Face this fact. All people, by their most base natures, truly love to belong to exclusive groups or clubs. The more exclusive the better. We call this the "Gold/Silver/Bronze Mindset" and you should structure within your curriculum such groups to cater to this mindset in your pupils.

Additional curricular entities such as a Black Belts only class, or an Instructorship training program, or perhaps a special executive fighting training program, etc. are all good tools for tapping into this mindset. Make people feel special, because they truly are, and they will reward you with undying loyalty. And undying loyalty translates to retention, which then in turn will translate to continuous growth or both your pupils and your school.

Implement any or all of the strategies above as you see fit. Be creative, but not hokey or cliché. Be consistent but not heavy handed. Be impeccable and fair minded. Truly care about your pupils and their progress. Do all of these things well and you will be rewarded with long term enthusiasm and retention in your student base which will translate to long term success and growth in your school.

Again, remember, it's about the student. It is not about you, the teacher. We are there to serve them to the best of our ability. They, and their progress within our respective martial arts and disciplines, are our *report card.*

Business is War - Pick Your Battles Wisely

Just as in fighting, it is easy in our industry to develop a mindset that you must try and defend each and every attack or perceived attack. However, just as in fighting, some attacks need no defense at all. And moreover, by attempting to defend, you may merely give your opponent an opportunity to bridge and/or capitalize on the enjoining of combat. Or, at the very least, you will expend your energy and precious resources by engaging what does not need to be engaged.

In business, just as in fighting, sometimes the best defense is simply to yield and do nothing at all, at least for awhile. Allow your opponent to tire while attacking air. Let them expend *their* energy and resources while you grow stronger and wait for the right timing to make your counter move.

"Keep your Guard Up and Look for opportunity to Convert"

When your normal marketing strategies are not working as usual, it is of course essential that a good warrior evaluates why this has come to pass. But a time worn, tested and effective strategy must be given its due credit and not be discarded casually without proper evaluation. In certain instances, it may simply be the wrong timing for that strategy. Or, the marketing landscape may have changed slightly necessitating a minor adjustment before reimplementation.

See the big picture and the Annual Aggregate. Don't panic with monthly ups and downs.

All business is cyclical. The martial arts business is no different. The business trend can, and will, vary throughout the twelve months of each calendar year. You will also experience trends that correspond with certain, specific times of each year. The longer you are in business, the more experience you will gain in your specific local area and market regarding such trends. But, for all schools, the following trends are pretty much a given year in and year out and you should expect them, and moreover, anticipate and plan ways to balance the likely down trend periods and/or capitalize on the likely up trend periods.

January*– New Years Resolutions*

You will receive an influx of those folks who are feeling the effects of their perception of a "new year, new start", a recent Birthday, their recent life changes or simply their recent binge eating during the holidays. You should have a New Year's Special promotional package in place before you go on your own holiday finalized and set for implementation IMMEDIATELY after the New Year turns.

***July**– The Vacationers*

July is the most popular vacation month in the country. You can expect that this month will be a very slow time in terms of new enrollees. Offset the down trend of new enrollees with in-house seminars during the month of July itself to keep your revenues higher. You can also try running a promotional special in the preceding months of May and June to attempt to garner a few more new enrollees than usual to average out your influx of new students over the summer.

***December**– The Holiday Lulls*

This is also a tough period for new enrollees. For the most part, everyone stopping into your school will wish to "postpone" their enrollment until the craziness of the holiday season is over. Two strategies can be employed here effectively.

One) Maximize the "Gift Giving Season" through your pro shop sales. Offer special sales deals in conjunction with bulk purchases or pairings of specialty items that are not normally on sale throughout the rest of the year.

And two) Try offering a "sign now, pay in January" promotional special that gives some sort of concession to the enrollee willing to take you up on such an offer. Perhaps a free item or items that have substantial value to them, or, a free private lesson or two.

Stay vigilant. Keep your guard up and your steps fluid and lively.

Knowing when the highs and lows are likely to hit will help keep you focused on the overall goal. You will feel empowered by having that knowledge and be committed to helping your circumstance, as opposed to feeling powerless in the face of what are to be certain business "swings".

Never EVER find yourself "Crying poor during the Down Months" to ANYONE, including your staff and especially not your students. This is analogous to showing your opponent that you are hurt. This only signals to your students your weaknesses and may potentially communicate a lack of focus to those same individuals. Nobody feels comfortable when they begin to feel that they are associating themselves with a "Loser". You must take pains to never send this message in any way.

Also, remember not to get too complacent during the "Up Months" of your business trend. "What goes up must come down" and no up trend lasts forever. Knowing that the up trend won't last will keep you focused and proactive in analyzing why this may be occurring now and what parts of it you may be able to perpetuate for a longer duration. This perspective should also make you more mindful of those trends that may be "fad driven" and give you ideas of how you may balance those perspectives in your market share as the drive behind the fad inevitably begins to wane.

So remember... Guard up. Never be complacent and let it down in the face of the war. Watch for opportunity and capitalize when you see an opening. And understand that within the war, there will be those skirmishes lost, and those won. And one battle does not a war make.

Addendum One contains actual advertisements that were printed in local phone books, community newspapers, major supermarket circulars, etc. The ad sizes fluctuate based on budget and space availability. The ads chosen for this section were all very successful for us at the time. You can use the examples given as templates that can be modified, customized or changed for your next print advertisement if you so choose.

Page 127:

This high cost, half page ad generated a great deal of traffic into our school and we ran it for several years. The larger ad space allows for photos and greater content in the message. We ran this ad after we had maximized our facility, expanding to occupy the entire ground floor of our building. It was a successful ad campaign which brought us many new members. However, it is the content of the advertisement that makes the impact. You must balance size with budget as you go along and remain mindful of this fact.

Page 128:

This 1/4 page ad allowed us to diversify our budget so that we could also designate a portion of our budget to internet advertising. The slightly smaller ad space delivered our message just as well. It was the content and the appearance of the ad that captured the attention of our potential students. The ad primarily directed interested individuals to our extensive, information based website which we had recently brought into being. This marketing strategy generated substantial traffic to our schools and over time actually yielded a higher sign-up rate in comparison to the high cost 1/2 page ad we had run prior.

Page 129:

We ran this modest size ad in our local supermarkets' weekly circular to promote the softer arts in our curriculum. Given that the market share of the softer arts was considerably less than our primary curriculum initially, this smaller sized ad allowed us to stay within our budget and still get the message out regarding these other options of study in our school.

San Diego's Best Kept Secret!

Since 1978 - Voted One of the Top Ten Kung Fu Schools In the Nation

Chinese Internal Kung Fu Systems

Hsing I - Pa Kua - Tai Chi

3 Schools In One! Spacious Traditional Three Training Hall Facility!

*Master ——————- 31yrs. Exp. 2*** —— International Champion. USA ——- Member of Hall of Fame. Nationally recognized Master Instructor of —————————.*

No Other School Can Compete With Our Quality Programs, Lifelong Experience of Our Top Teachers and the Personal Attention Given to Our students!

Call Now For Peace of Mind

200-123-4000

www.****.com**

"Awareness & Confidence are the cornerstones of personal freedom."

- **Adults Only Institution.**
- **Exclusive Membership.**
- **Increase Energy and Focus through Traditional Martial Study and Meditation!**

*Instructor —————— 20 yrs. Exp. U.S.A. 2*** ——- Grand Champion. International ———- Chinese Martial Arts ——— ————— ————.*

Traditional training promotes the skills necessary to cope with the modern world. Come train with us in a proven teaching structure.

Your School Name Here

***** Anyway Dr. Your City, CA 90000**

Simply The Best!

Chinese Internal Kung Fu

Hsing I - Pa Kua - Tai Chi

Since 1978 - Voted One of the Top Ten Kung Fu Schools In the Nation

Master ————— 31yrs. Exp.
*2*** —— International Champion.*
USA ——— Member of Hall of Fame. Nationally recognized Master Instructor of—— —————.

Call Now For Peace of Mind

200-123-4000

- **Adults Only Institution.**
- **Exclusive Membership.**
- **3 Schools in One! Spacious, Fully Equipped, Traditional Three Training Hall Facility!**

Visit Our Website for more Information

www.****.com**

"No Other School Can Compete With Our Quality Programs, Lifelong Experience of Our Top Teachers, and the Personal Attention Given to Our students!"

Instructor ————— 20 yrs. Exp.
*U.S.A. 2*** ——— Grand Champion.*
International ——— Chinese Martial Arts ——— ———— ———-.

Your School Name Here

***** Anyway, Your City, CA 90000**

Are you Stressed Out?
Are you sick and tired of being SICK and TIRED?

Tai Chi Classes

Tai Chi has been Medically Proven to Reduce Stress and Raise the Body Immunity Levels.

Your School Name Here

Teaching La Mesa Since 1978.
Voted One of the Top Ten Schools in the Nation!
Trained in Taiwan.
IKF Hall Of Fame Coach of the Year!

Introductory Offer: **Train as a couple, enroll with your spouse today!**

www.********.com

*** Anyway Drive; Your City, CA. 90000

CALL TODAY to get started! (200) 123-4000

Addendum Two covers promotional packages with which we have had great success over the years. They may be useful to you as ideas to be modified to fit your school. You can customize them as desired by inserting your own school logo, curriculum, price point, retail costs of any merchandise, expiration date, vital information, etc.

<u>Page 131</u>: This has been one of the most successful promotional packages we have ever offered. The package includes everything any new student would need in the near term of their studies at a tremendously discounted rate. They are being rewarded with such a package when enrolling in a long term program. We experienced an 80% sign up rate with new enrollees choosing this package over all others presented.

<u>Page 132</u>: This promotional package is intended as interim example for those operations that may be in transition from programs that are "Group Classes *only"* to what we recommend as optimal "Group and Privates Classes" combined program offerings in their schools.

<u>Page 133</u>: You might consider offering this short term program to those prospects who are not ready to commit to long term studies. Or, during the summer vacation period for those prospects who are committed to academic pursuits the rest of the year.

<u>Page 134</u>: You might offer this ever popular, all inclusive promotional package during the end of, or the beginning of the year, for those who make New Year's Resolutions.

<u>Page 135</u>: This promotional package is perfect for those individuals who want to jump into their martial arts pursuit with both feet. Be sure that you offer the necessary corresponding sparring classes that *all* levels of students are eligible to attend.

<u>Page 136 and 137</u>: Both of these promotional packages are popular among individuals without prior martial arts experience who seek study for health benefits. A short term program is especially palatable for them to get involved before committing to long term study. Notice this package does not include workout attire or a uniform. This will present an opportunity for "up sell". Be creative.

<u>Page 138</u>: This promotion is specifically designed to entice new members by direct advertising. This type of "Door Hanger" can be utilized for grand openings and/or occasional door-to-door advertisement to generate awareness in your local residential areas. Don't do deep discounts often.

<u>Page 139</u>: Guest Passes can generate new members brought in by other students simply because friends enjoy sharing positive experiences with other friends. Hand them out to new pupils.

PATH TO MASTERY STUDIES

(For the more committed, long term student.)
A $364.00 Instructional package for

ONLY $189.00

Your *Path To Mastery Course Includes* ***all*** *you need to get started right...*

• 1 month of Group Classes. (no limit attendance, come every day)	normally $ 140
• 1 Introductory Private Lesson.	normally $ 60
• Association Fee Included. (registers you internationally)	normally $ 45
• 1 Free Uniform. (top & bottom included)	normally $ 45
• 1 Complete set of School Patches plus training shoes	normally $ 26
• 1 Free School T-Shirt (your school t-shirt)	normally $ 18
• 1 Beginner's Video Tape or DVD. (Featuring all required forms & Applications.)	normally $ 30

All of this is only $*189.00* when you sign up as a new student under any of our regular student plans. $*189.00 is all you pay for this entire one month Path To Mastery course.* Your regular plan will commence immediately after your *Path to Mastery* course ends. What a great beginning!

Learn: (insert your own curriculum below)

⇒ Self Defense, Greater Control and Peace Of Mind
⇒ Tai Chi and Chi Kung for Stress Reduction
⇒ Hsing I, Tai Chi and Pa Kua for Better Levels of Energy
⇒ Meditation, Philosophy and Chinese Medicine

Your first step toward greater peace of mind should be taken Today!
With all we have to offer, you will be happy that you did!

CALL TODAY! (expiration date here)

The longest journey starts with a single step, so START WALKING!

Your School Name Here
***** Anyway Drive**
Your City, CA 90000
(200) 123-4000
www.**.com**

25th Anniversary Intro Special

(Short Term Commitment; No Further Obligation.)
A $376.00 Instructional package for

ONLY $295.00

Your *Trial Study Course Includes* *<u>all</u>* *of the following:*

- 2 mos. Group Classes. (no limit attendance, come every day) — normally $ 260
- Association Fee Included. (registers you internationally) — normally $ 45
- 1 Full Uniform. (top & bottom included) — normally $ 45
- 1 complete set of school patches and training shoes. — normally $ 26

This Trial Study course is yours for only *$295.00* when you enroll as a new student on a trial basis. *$295.00 is all you pay for this entire trial study course.* You are under no further obligation to continue your studies after your Trial Study course expires. Although we think that you will wish to do so!

<u>Learn:</u> (insert your own curriculum below)

⇒ Self Defense, Greater Control and Peace Of Mind
⇒ Tai Chi and Chi Kung for Stress Reduction
⇒ Hsing I and Tai Chi for Better Levels of Energy
⇒ Meditation, Philosophy and Chinese Medicine

Your first step toward greater peace of mind should be taken Today!
With all we have to offer, you won't be sorry you did!

CALL TODAY! (expiration date here)

Your School Name Here
***** Anyway Drive**
Your City, CA 90000
(200) 123-4000
www.**.com**

Get Started All-ready Intro Course

(Short Term Commitment; No Further Obligation.)
A $516.00 Instructional package for

ONLY $369.00

Your *Trial Study Course Includes* ***all*** *of the following:*

• 2 mos. Group Classes. (no limit attendance, come every day)	normally $ 280
• 2 Introductory Private Lessons.	normally $ 120
• Association Fee Included. (registers you internationally)	normally $ 45
• 1 Full Uniform. (top & bottom)	normally $ 45
• 1 complete set of school patches and training shoes.	normally $ 26

This Trial Study course is yours for only *$369.00* when you enroll as a new student on a trial basis. *$369.00 is all you pay for this entire trial study course.* You are under no further obligation to continue your studies after your introductory course expires. Although we think that you will wish to do so!

<u>Learn:</u> (insert your own curriculum below)

⇒ Self Defense, Greater Control and Peace Of Mind
⇒ Tai Chi and Chi Kung for Stress Reduction
⇒ Hsing I and Tai Chi for Better Levels of Energy
⇒ Meditation, Philosophy and Chinese Medicine

Your first step toward greater peace of mind should be taken Today!
With all we have to offer, you won't be sorry you did!

CALL TODAY! (expiration date here)

Your School Name Here
***** Anyway Drive**
Your City, CA 90000
(200) 123-4000
www.**.com**

Get Started All-ready Intro Course

(Short Term Commitment; No Further Obligation.)
A $516.00 Instructional package for

ONLY $369.00

***Your** Trial Study Course Includes <u>all</u> of the following:*

• 2 mos. Group Classes. (no limit attendance, come every day)	normally $ 280
• 2 Introductory Private Lessons.	normally $ 120
• Association Fee Included. (registers you internationally)	normally $ 45
• 1 Full Uniform. (top & bottom)	normally $ 45
• 1 complete set of school patches and training shoes.	normally $ 26

This Trial Study course is yours for only *$369.00* when you enroll as a new student on a trial basis. *$369.00 is all you pay for this entire trial study course.* You are under no further obligation to continue your studies after your introductory course expires. Although we think that you will wish to do so!

<u>*Learn:*</u>

⇒ Self Defense, Greater Control and Peace Of Mind
⇒ Tai Chi and Chi Kung for Stress Reduction
⇒ Hsing I and Tai Chi for Better Levels of Energy
⇒ Meditation, Philosophy and Chinese Medicine

Your first step toward greater peace of mind should be taken Today!
With all we have to offer, you won't be sorry you did!

CALL TODAY! (expiration date here)

****** Anyway Drive Your City, CA 90000**

(200) 123-4000

"COMBAT" TRIAL COURSE

(Short Term Commitment; No Further Obligation.)

A $546.00 Instructional package for

ONLY $449.00

Your *"RSPCT" Trial Course Includes* ***all*** *of the following:*

- 2 mos. Group Classes. (no limit attendance, come every day.) normally $ 280
- Association Fee Included. (registers you internationally) normally $ 45
- 1 Workout Uniform. (shirt, pants & shoes) normally $ 41
- 1 complete set of "Combat" Gear. (regulation quality)
  set includes: Head Gear, Rib Guard, Gloves, Shin & Instep Guard, Cup Support normally $ 180

This "Combat" Trial course is yours for only *$449.00* when you enroll as a new student on a trial basis. *$449.00 is all you pay for this entire trial study course.* You are under no further obligation to continue your studies after your "Combat" Trial course expires. Although we think that you will wish to do so!

Learn: (insert your own curriculum below)

⇒ To Take Fitness to the Next Level
⇒ Real World Defensive Skills in a Graduated Method
⇒ No Nonsense Techniques through a Proven Approach
⇒ Self Empowerment through Martial Arts

Your first step toward a Greater Fitness Level should be taken Today!
With all we have to offer, you won't be sorry you did!

CALL TODAY! (expiration date here)

Your School Name Here
***** Anyway Drive**
Your City, CA 90000
(200) 123-4000
www.**.com**

"CHI KUNG" COURSE

A Short Course In Energy Development
A $300.00 Instructional package for

ONLY $149!!!

***Your** Chi Kung Course Includes all of the following:*

- 2 mos. Group Classes! (limited to Tai Chi & Meditation classes only.) normally $ 180
- Two Private Lessons, One on One, for 30 minutes each. normally $ 120

This Chi Kung course is yours for only *$149.00* when you enroll as a new student. *$149.00 is all you pay for this entire two month Chi Kung course!* You are under no obligation to continue your studies after your Chi Kung course expires. Although we think you will when you find out how enjoyable it is!

Learn: (insert your own curriculum below)

⇒ Greater Control and Peace Of Mind
⇒ Chi Kung is excellent for Strengthening the Immunity systems.
⇒ Discover Better Levels of Energy within.
⇒ Practice Meditation for strength of mind.

Your first step toward greater peace of mind should be taken Today!
With all we have to offer, you won't be sorry you did!

CALL TODAY! (expiration date here)

Association Fee Is Not Required For This Course.

Your School Name Here
***** Anyway Drive**
Your City, CA 90000
(200) 123-4000
www.**.com**

"TAI CHI" COURSE

Learn The Tai Chi Short Form Of Practice
A $300.00 Instructional package for

ONLY $149!!!

***Your** Tai Chi Course Includes all you need to get started right!!!*

- 2 mos. Group Classes! (limited to Tai Chi & Meditation classes only) normally $180
- Two Private Lessons, One on One, for 30 minutes each. normally $120

All of this is yours for only *$149.00* when you sign up as a new student in our Tai Chi Program. *$149.00 is all you pay for this entire two month Tai Chi course!* You are under no obligation to continue your studies after your Tai Chi course expires. Although we think you will when you find out how beneficial it is!

Learn: (insert your own curriculum below)

⇒ Greater Control and Peace Of Mind.
⇒ Tai Chi and Chi Kung for Stress Reduction
⇒ Discover Better Levels of Energy within.
⇒ Meditation, Philosophy and Chinese Medicine.

Your first step toward greater peace of mind should be taken Today!
With all we have to offer, you won't be sorry you did!

CALL TODAY! (expiration date here)

Association Fee Is Not Required For This Course.

Your School Name Here
***** Anyway Drive**
Your City, CA 90000
(200) 123-4000

DOOR
HANGER

YOUR SCHOOL NAME HERE
GRAND OPENING SPECIAL!!!
CELEBRATING THE NEW YOUR CITY BRANCH!
A $340 package for

ONLY $139!!!

Your *Introductory Course Includes:*

• 1 mos. Group Classes!	normally $140
• Association Fee Included!	normally $ 45
• 2 Private lessons! 30 minutes each	normally $120
• 1 Basic Forms Video Tape!	normally $ 30
• 1 Free Uniform!	normally $ 45

All of this is yours for only *$139.00* when you sign up as a new student under any of our regular student plans. *$139.00 is all you pay for the first month that you study!* Your regular plan will commmence after your Introductory course expires!

Learn: (Insert your school curriculum below.)

⇒ Self Defense, Greater Control and Peace Of Mind
⇒ Tai Chi and Chi Kung for Stress Reduction
⇒ Hsing I and Pa Kua for Better Levels of Energy
⇒ Meditation, Philosophy and Chinese Medicine

Superior Group or Private Instruction for Men and Women. Train at either or both of our facilities at your convenience. Classes Monday through Saturday at both locations throughout the day. Open 10 to 9 at both locations.

Be a part of our 15 Year Anniversary Celebration Today!
With all we have to offer, you won't be sorry you did!

***Anyway, Your City, CA 90000
On the corner of Anyway and 1st Street, in XYZ Shopping Mall

CALL TODAY

(200)123-4000

www.****.com

Your School Name Here

V.I.P. GUEST PASS

Participate in one of the following types of Classes absolutely FREE:

- Hsing I for Self Defense and High Energy workouts
- Tai Chi for Stress Reduction and fitness of a softer nature
- Pa Kua for greater understanding of internal dynamics

Attend one free class at either our New Branch or the Main Branch, 5,000 square foot mother School. We are the largest Internal Kung Fu Organization on the West Coast, trainer of numerous International, National and Regional Internal Kung Fu Champions. We have the most modern, well equipped and clean facilities in San Diego County! Check It Out!

This invitation is good only upon confirmation through the office. Please call to schedule your free class ahead of time.

Your School Name Here
***** Anyway Drive**
Your City, CA 90000
(200) 123-4000

Addendum Three covers advertisements for seminars and/or workshops that have been offered in our schools. All were well received. Always be certain to include a deadline of "no-refund" for those seminars or workshops that include *free gifts*. It is our experience that the deadline for refund should be *before* the day you place the order for such gift items. This will prevent excess inventory burdens in your retail shop.

Page 141 and 142: These are examples of possible "softer sides" of studies. You can keep the cost low in such venues because there is no merchandise offered in such events. You should consider including a printed hand-out of the material to be covered to all attendees. The cost is paper, ink and your time to prepare, but the perception of value is delivered. The attendees will now have their own reference to further their studies. Without the hand-out, they probably will not be able to retain the information and thus will likely not attend the next event.

Page 143: This workshop creates a context of sequential and related events so that your students are able to perceive the continuity of such studies. Knowing what is coming breeds excitement in your students generated from the anticipation of the next upcoming block. This type of sequenced event could also act as a spring board into new classes added to your curriculum in the near term. New classes stimulate enthusiasm within your student body. An additional side benefit of this type of workshop is the stimulation of training equipment sales by virtue of making such equipment a requirement for ALL attendees.

Page 144: This is an example of a very successful fundraising event for us at the time. Often, you can use such events to raise funds for certain charities and/or utilize the event to generate awareness in your local community.

Pages 145 through 150: These are seminars and/or workshops which were highly successful in our school. The word of mouth carried throughout our school after the first event. We experienced an overwhelming participation in these events among our student body. The responses were so positive that we had to run repeats of the same workshops shortly after the initial event. These events generated a great deal of enthusiasm toward classical weaponry studies.

Chi Kung Workshop
The 36 Longevity Exercises

October 15 10:00a.m. to 12:00pm

<u>Subjects Covered</u>:

The Eight Preparations

The Fourteen Passive Stimulators

<u>Subjects Covered</u>:

The Fourteen Active Stimulators

Seated Meditation

Shr Fu will be teaching the Taoist 36 Longevity Exercises, their methods of practice and the associated breathing techniques.

This is an ancient series of excercises containing components to Calm the Mind, Center the Breath, Retard Aging and Stimulate Key Systems in the body. These are simple exercises that can be performed in about 20 minutes each day for long term health maintenance.

He will also be teaching "Sit Chen" or Seated meditation that same day and discussing the basic goals and methods of both energetic practices and Taoist introspective thought. The fee will be $50.00 for the day.

Don't Miss It!

Eh Mei Shen Shu Seminar – 3/15 10 am to 12 pm

Your School Name Here

The Eh Mei Shen Shu (Mountain Spirit Method) trainings go all the way back to the beginning of the Taoist traditions for balancing and restoring energy levels in the male body which can be compromised over time through regular and normal sexual activity. Relatively simple to learn, they can make a substantial difference through a lifetime of practice. This body of exercises are male gender specific (if sufficient interest exists for the female population, a similar seminar can be done in the future for women).

The seminar will focus on the Eh Mei trainings in addition to general information regarding the traditional Chinese medical view of male essence and sexual alchemical teachings of conversion and restraint.

The fee is $60 for the day. *Fees must be in no later than March 7th.* ***(No Refunds will be given after March 7th.)***

This seminar is Open to All Levels of male students, regardless of whether they have practiced chi kung before or not. The seminar will be limited in number of participants.

If interested, notify the office immediately to reserve your spot. It is filling quickly..

Sparring Tactics Workshops 20XX-XX

Your school name here

Bridging Tactics

Date: 9/12/20XX

Time:

Defending The Bridge

Date: 12/12/20XX

Time:

Breaching The Gate

Date: 3/12/20XX

Time:

I have decided to begin holding a series of Sparring Tactics Workshops throughout the calendar year. And if well received, this will likely become a staple of the curriculum in addition to other regular classes and seminars.

I have always believed that the combative aspects of the martial arts need to be integrated as part of the total scale of training. It has become apparent to me that many people do not want to make a habit of free sparring on a regular basis. However, this does not preclude that many may want to learn about the tactics therein. In talking with many of the students, it seems that there are a variety of factors that prevent regular attendance of many within the scheduled Wednesday night combative class. So, the purpose of these workshops will be to introduce key concepts of tactical combat in a forum devoid of free sparring in the conventional sense. This will allow students to explore and learn about combat perspectives in controlled two person activities while eliminating all risk of the unknown.

We plan to run these workshops as a series of topics one leading into, and building on top of, the next in succession.

Each workshop will be offered at a cost of $45 which is due and payable upon sign-up.

All attendees to these workshops must have purchased the protective gear of: Masks, Gloves and Shin/Instep protectors. Other protective gear will not be necessary but may be worn if desired.

All workshops will be available to students of all levels. However, less senior students will need to attend the first in the series and then progress from there. The more senior students may choose to come selectively if desired.

It is my hope that these workshops may offer an alternative method of learning and developing some of these skills in lieu of the combative class. I strongly recommend this exposure.

Highlights

- Distance and Insertion Skills
- Folding and the Seven Stars
- The Nine Gates Theory

Develop New Combative Skills or

Enhance Existing Skills by developing Greater Insight...

TACTICS WORKSHOP

On ***Saturday, September 28th, at12 noon,*** *the United States Team Members will be hosting a* ***workshop to teach their favorite fighting techniques and strategies*** *in an effort to raise the overall skill levels of upcoming students interested in the tactical side of martial arts training.*

The workshop will also be accompanied by an ***outdoor bake sale,*** *and the* ***proceeds from both the Tactics Workshop and the Bake Sale will be going to the fund to send the United States Team to Taiwan.***

We hope that all of you will show up to either participate or simply watch the festivities and enjoy the baked goods offered at the sale. Some of the baked goods are rumored to contain ***Chi building properties*** *in addition to a savory taste.*

Regular classes will be pre-empted on that Saturday.
Hope to see you there!

The fee for the workshop will be ***Twenty Dollars****. Payable in* ***cash only please*** *(and we will write you a receipt for the charity donation).*

OPEN TO MEMBERS ONLY!

Five Element Spear Seminar – 3/12 6:30 pm to 8:30 pm

Your school name here

This is planned as a two hour event covering the Hsing-I Long Spear routines (forms only). The Hsing I Spear routines are based on the five elements and, aside from having easily practical fighting applications, have long been practiced to strengthen core movement mechanics. The continued practice of issuing energy through an 80 inch spear will strengthen the body in ways that hand forms simply cannot do. These structures are beneficial at any level and the knowledge is fast becoming more rarefied. You must have a *traditional 80 inch waxwood* practice spear to attend this workshop. The office can provide one for you at a fee of $75.

Fees for spear order must be in no later than date (one week prior). ***(No Refunds will be given March 5th).*** *(when the spear order is placed)*

This workshop is open to All Levels of students, regardless of whether they have practiced spear before or not. Because of the length of the weapon and scope of the work, the workshop will be very limited in number of participants.

If interested, notify the office immediately to reserve your spot. It is filling quickly..

Stick Training Seminar – 9/12 10 am to 3 pm

Your school name here

This is planned as an all day event covering the Single Siniwali, the Double Siniwali and Reverse Siniwali, stepping patterns and tactics of stick fighting, plus selected self defense applications of the stick. Look at it as Stick Combat 101. Two pairs of Escrima sticks (one hardwood for the drills and one foam covered for the applications) will be included in the Seminar Cost (yours to keep). As well a break for lunch (Chinese). The fee is $95 for the day. This includes both sets of sticks and lunch costs.

Fees must be in by date. No refunds after September 5th.

This seminar is Open to All Levels of students, regardless of whether they have practiced with sticks before or not. These are classic training methods of the short stick. The utility of the stick as the great equalizer for self defense is widely known and appreciated. The size of the seminar will be limited to 40 students.

This seminar will be a pre-requisite for Stick Training Level 2, planned early next year. If interested, notify the office immediately to reserve your spot. It is filling quickly.

Your school name here

******Anyway, Your City, CA 90000***

Date: 8/15

Time: 10:00 a.m. to 3:00 p.m.

Sword Combat Seminar– August 15

This seminar is planned as an all day event to delve into numerous applications, concepts and tactics of swordplay. The material covered will be drawn from all three classic internal art sword methods but we will not be covering form at all that day. All students who have attended any of the sword seminars in the past and/or learned the Hsing I, Pa Kua or Tai Chi Jen (sword) methods during regular weaponry class are eligible to attend. A wooden practice sword/jen (yours to keep) and lunch (Chinese) are both included in the cost for the day.

This seminar will run from 10:00 a.m. to 3:00 p.m. with a one hour break for lunch. The fee for the day will be $80, which is due upon sign-up. No refunds after August 8th will be permitted.

Pa Kua Jen (Sword) Seminar –11/8 10 am to 3 pm

Your school name here

This is planned as an all day event covering the Pa Kua single hand straight sword routine. This is a strikingly beautiful and intricate straight sword routine for the sword aficionado. A beautiful stainless steel Lion Head sword with leather scabbard (see the display in the middle hall) and lunch for the day (Chinese) are included in the cost. The fee is $125 for the day. This includes the sword and lunch costs. *Fees must be in no later than 11/1.* ***(No Refunds after November 1st.)***

This seminar is Open to All Levels of students, regardless of whether they have practiced sword before or not. This is a beautiful, classic Pa Kua single hand straight sword routine for practice. Because of the scope of the work, applications for this sword routine will be planned for a future date. The size of the seminar will be limited to 40 students.

If interested, notify the office immediately to reserve your spot. It is nearly full.

Actual Sword and Scabbard on display in the middle hall.

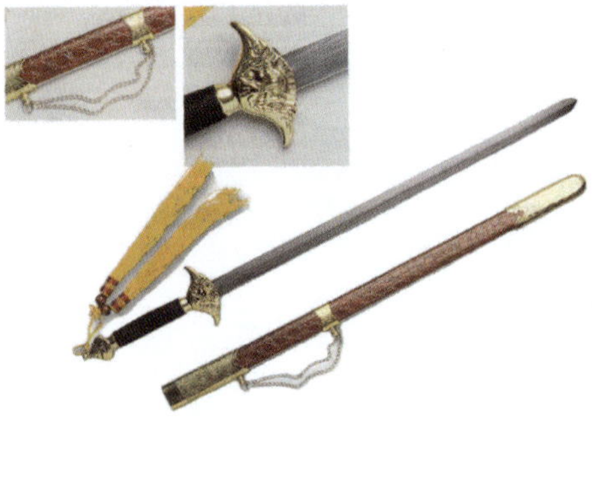

Two Hand Sword Seminar – 7/22 10 am to 3 pm

Your school name here

This is planned as an all day event covering the Hsing-I Two Handed Sword 5 Element Forms, Linked Form and Two person Combat Form plus selected applications from the singular forms. Look at it as Sword Combat 101. A Wooden Practice Sword will be included in the Seminar Cost (yours to keep). As well a break for lunch (Chinese). The fee is $75 for the day. This includes the sword and lunch costs. *Fees must be in no later than July 15th.* ***(No Refunds after July 15th.)***

This seminar is Open to All Levels of students, regardless of whether they have practiced sword before or not. This is an old classic Hsing-I training method to develop stronger core power using the weapon as a medium for that purpose. The mechanics incorporated in the practice are applicable to all internal arts (and your golf game as well). The size of the seminar will be limited to 40 students.

If interested, notify the office immediately to reserve your spot. It is nearly full.

Hsing I Jen (Sword) Seminar – 2/17 10 am to 3 pm

Your school name here

This is planned as an all day event covering the Hsing-I single hand straight sword routine and its applications. This is a wonderful and relatively simple sword routine to learn and practice. A beautiful wooden practice sword (yours to keep) and lunch (Chinese) are included in the cost. The fee is $85 for the day. This includes the sword and lunch costs. *Fees must be in no later than February 10th.* ***(No Refunds after February 10th.)***

This seminar is Open to All Levels of students, regardless of whether they have practiced sword before or not. This is an old classic Hsing-I single hand straight sword routine for practical use. The size of the seminar will be limited to 40 students.

If interested, notify the office immediately to reserve your spot. It is nearly full.

Addendum Four covers additional promotional campaigns to entice new members. Such events can be useful tools for broadening your market share. They can be modified to suit your curriculum, then printed and posted on your bulletin board to garner interest.

Page 152: An annual School Tournament can be a wonderful opportunity for interested students to be able to showcase their performance to their peers, personal friends and families. You might consider allowing your existing students to bring guests to this event. The witnessed enthusiasm and dedication of your students *will* encourage some of the guests to enroll in what they see as your school's outstanding curriculum.

Page 153: The catchy name of this event captured a great deal of interest in our students. So much so that they requested to bring along those friends whom had always been drawn to the holistic way of life. Such events can truly showcase the depth and extensive nature of your curriculum.

Page 154: This is yet another example of holistic studies. And the possibilities of a diversified curriculum within your school.

Annual School

Tournament

BE A PART OF 20XX EVENT

- Forms Competition
- Weapon Forms Competition
- Push Hands Competition
- Sparring Competition
- Weapon Sparring Competition

Friends and families are invited to this fun, exciting, inspirational event. Space is limited. RSVP–A.S.A.P.

SEPTEMBER 8TH
6:30 P.M.
THE MAIN HALL

Free Exercise for those who would like to train that evening.

Vitalithon

Saturday - January 20th - 10 a.m. to 12 p.m.

Subject Covered:

**In Depth:
Five Element
Chi Kung
Internal Practice**

Subject Covered:

**How the Five
Elements are
used in TCM
diagnosis and
Acupuncture
Treatments**

All Students MAY BRING A GUEST to this FREE SEMINAR/ LECTURE focused on learning about the Five Energies of Traditional Chinese Medicine and Internal Martial Arts Practice for Health and Vitality.

If you wish to Bring a Guest, please R.S.V.P. through the office and give the name of the guest as well. All guests will be required to wear a pair of our practice kung fu shoes to be allowed on our practice floor, so purchase them a pair before the date arrives. All guest will be required to arrive 15 minutes early and fill out a waiver form to be allowed on our practice floor. These two requirements MUST be met for each guest and both requirements will be your responsibility to ensure their participation.

Don't Miss It!

The Tao of Conservation And Transformation

Saturday - March 1st - 10 a.m. to 12 p.m.

All students are allowed (and encouraged) to bring a guest to this event.

On the above date and time the Institute will be hosting a free Seminar on the Concepts of Energy Conservation and Transformation according to Taoist principles as found in the Shen Lung Internal Martial Arts. This seminar will be composed of a brief lecture (on the subjects of food combination and energy conservation/transformation), a participatory introduction to the Five Animal Frolics (we will be performing the Tiger, Bear and Bird exercises), and then the floor will be thrown open to questions until time expires.

If you wish to Bring a Guest, please R.S.V.P. through the office and give the name of the guest as well. All guests will be required to wear a pair of our practice kung fu shoes to be allowed on our practice floor, so purchase them a pair before the date arrives. All guest will be required to arrive 15 minutes early and fill out a waiver form to be allowed on our practice floor. These two requirements MUST be met for each guest and both requirements will be your responsibility to ensure their participation.

Don't Miss It!

Addendum Five covers specific signs that were used in our schools. These computer produced, posted signs demonstrate professionalism within your operation. You may consider modifying any of them to suit your purpose. Consider laminating those postings that are to remain on the wall for a long period of time. Torn edges or unsightly signs represent quite the opposite of a professional atmosphere.

Page 156: ***Insurance Regulations Regarding Jewelry***

This potential safety/liability hazard is often overlooked by school operators and instructors. Injuries and property damage could send a negative message to all parties involved. Laminate this notice and post in the student's lounge and training areas so that *All* members are informed.

Page157: ***Check Writing Policy***

A check writing policy should be clearly visible in the office and preferably laminated. Post pertinent information for those who write checks and eliminate time spent in answering basic questions. Repetitive questions reduce your productivity.

Page 158: ***Closed For Repair***

Use this kind of sign when repairs, routine necessary maintenance, renovations or remodeling are happening in your school. Provide information about how long the inconvenience will affect your operation. If necessary, warn people about any safety issues.

Page 159: ***New Classes Forming***

Post such a sign on the bulletin board or student lounge whenever such things occur.

Page 160: ***Right To Refuse Service***

Post such a sign in a conspicuous place for *All* to see.

Page 161: ***Remove Your Street Shoes***

This may not apply to all facilities. Post this sign at the entrance of any area in which you do not wish students to track dirt into your training floor or other parts of your facility.

Page 162and 163: ***Change of Season, Change of Uniform***

Post such a sign on the bulletin board at least 2 weeks or more before the deadline. This notice will generate sales in your retail shop twice per year.

Page 164: ***Retail Shop Price List***

Laminate such a sign and post in your retail corner or pro-shop.

Price adjustment strategy: Add additional 10% to the regular retail price for all items sold to non-members. The 10% discounted price becomes a perk to *ALL* students. Membership has its privileges.

NAME OF YOUR SCHOOL HERE

To All Students – Today's Date

Insurance Regulations Regarding Jewelry

Recently, the insurance regulations governing martial arts training have been updated with regard to wearing jewelry while engaged in training. The basic language is as follows:

No jewelry of any kind is to be worn while engaged in training in any martial art. This includes, but is not limited to:

Watches, Rings, Earrings, Necklaces, Pins and "Piercing Jewelry" of any sort. The wearing of jewelry while engaged in training constitutes a potential safety hazard. It is an unacceptable risk.

Please obey the guidelines. Not doing so puts yourself, your partners and the school at risk of injury liability.

NAME OF YOUR SCHOOL HERE

*** Anyway, Your City, CA 90000
(200) 123-4000

Check Writing Policy
Minimum Amount $25.00

Credit Card Policy
Minimum "Charge" $50.00

All Checks must be made payable to:

Your School Name Here

Returned Checks will be subject to a $25.00 fee for bank costs and the two week delay in actual collection of the funds.

There will be absolutely NO CASH REFUNDS on any Credit Card Purchases.

10/01/XX

Closed For Electrical Repair

The Institute will need to be closed on this Friday, the 1st, through Sunday, the third of December, for routine electrical repairs. The workmen will need to access the facility at will and their equipment will be strewn all over in less than predictable fashions. Therefore, we have elected to close for this Friday night's class to avoid a safety hazard. The work will be complete within three days and we will open back up again for regular classes on Monday, the 4th, as usual.

Sorry for any inconvenience this may cause anyone.

R.S.P.C.T.

The idea of this program was spawned from the long years of involvement in training students to compete in the KuoShu structure and is the product of over two years of planning since that involvement. Many people nowadays wish to learn the essence of fighting and strategy therein without investing the time to learn an entire traditional martial system. This program will allow just that.

There will be no conventional ranking in this program. Students will advance solely on the merit of practical combative skills. The training will be implemented through conceptual blocks of knowledge. Testing will be in the form of scenario matches to see if the student is able to implement each level of tactical training and strategy. This program can be separate from, or in addition to, traditional training at the institute.

Tuesdays & Wednesdays

Please Note . .

Anyone attending this class must have purchased a full set of training gear through the institute so that we can satisfy existing insurance regulations. Gear purchased outside the institute (even if the same kind) will not be acceptable due to our requirements of reporting.

We Reserve The Right To Refuse Service To Anyone.

Posted by the Management
Name of your school here

PLEASE REMOVE SHOES BEFORE ENTERING PREMISES!

Please Do Not Leave Them Outside.
Thank You.

Wearing outside shoes into the school will track dirt, debris and grease on to the practice floor. Please endeaavor to keep the facility clean. Failure to do so will result in the spiritual conditioning of 200 push ups for the student member. Visitors, please stay along the outside of the floor! DO NOT walk on to the floor.

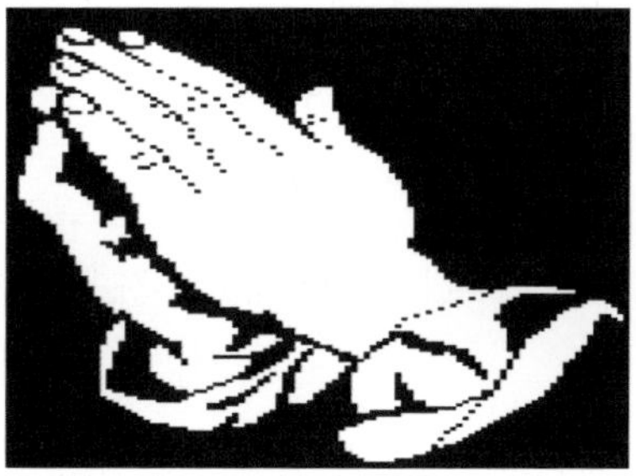

We appreciate the respect you show our school.
And, the respect you show us as human beings.

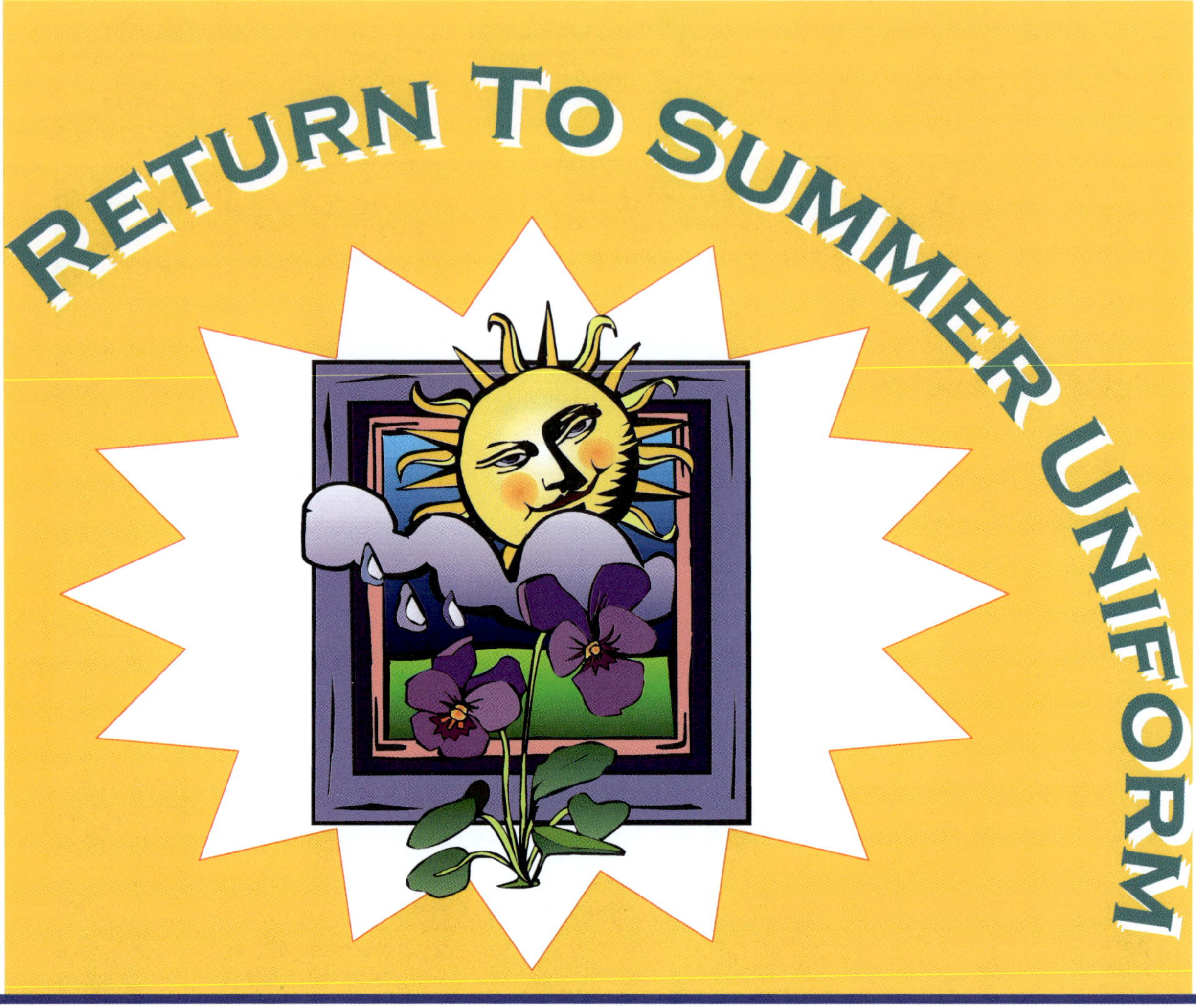

EARLY SUMMER?

It seems that the hot months are rapidly approaching us. We will be going to the Summer Uniform as of April 2nd, so that all may be more comfortable when training. If it continues to stay so warm, we shall soon be turning on the A.C. as well.

The temperature may continue to fluctuate a bit so feel free to wear a Gi top when you feel cold or a T-Shirt when you feel warm.

Please remember the summer uniform is an Institute T-Shirt with regular Gi pants and appropriate rank insignia.

Return to Winter Uniform

End Of October?

It looks like Fall has started to arrive in terms of temperature. We'll go ahead and make the change back to Winter Uniform as of the end of October, on the 28th.

Please remember that this means a complete uniform with the appropriate colored Gi top and the appropriate patches in place, etc.

As always, feel free to wear your School T-Shirt under your uniform if too cold or the T-Shirt by itself if too warm. And be sure to spend a little extra warm-up time before training.

Name of your school here
PRICE LIST

		RETAIL	Student DISCOUNT
UNIFORMS:			
Chinese Style	size 3,4	$42.00	$37.80
	size 5,6,7	$44.00	$39.60
Japanese Gi	size 3,4	$39.00	$35.10
	size 5,6,7	$41.00	$36.90
Pants Only	All Sizes	$17.00	$15.30
Cotton Shoes	All Sizes	$ 7.00	$ 6.30
Rank Sashes	All Colors	$35.00	
Dragon Patch	(separately)	$ 9.00	
Taiwan Flag	(separately)	$ 5.00	
The Above Two As A Set		$12.00	
T-SHIRTS:			
12 Animals T-Shirt		$16.95	
SPARRING EQUIPMENT:			
Full Cage Headguard		$67.00	$60.30
Gloves (Fingerless)		$29.00 pair	$26.10
Rib Guard		$30.00	$27.00
Chest Guard		$39.00	$35.10
Shin/Instep Guards		$14.00 pair	$12.60
Knee Guards		$ 7.00 pair	$ 6.30
Elbow Guards		$ 7.00 pair	$ 6.30
Cup Supporter		$ 8.00	$ 7.20
TRAINING EQUIPMENT:			
Training Stick		$18.00	$16.20
Focus Mitt		$24.00	$21.60
Thai Kick Arm Shield		$24.00	$21.60
Foam Kicking Shield		$46.00	$41.40
Eagle Catcher		$16.00	$14.40
Tiger Claw Training Shoes		$46.00	$41.40
HERBAL SUPPLIES:			
Teh Ta Yao Bruise Liniment		$15.00	
Iron Palm Training Formula		$15.00	

NOTE: Student members receive a 10% discount on all uniforms and sparring equipment. (No additional discount applied to herbal supplies, video tapes, T-Shirts, Sashes or Patches.)

Printed in Dunstable, United Kingdom

91338443R00094